The Structure and
Evolution of Galaxies

The Structure and Evolution of Galaxies

Steven Phillipps

Astrophysics Group
Department of Physics
University of Bristol

John Wiley & Sons, Ltd

Other Wiley Editorial Offices

John Wiley & Sons Inc., 111 River Street, Hoboken, NJ 07030, USA

Jossey-Bass, 989 Market Street, San Francisco, CA 94103-1741, USA

Wiley-VCH Verlag GmbH, Boschstr. 12, D-69469 Weinheim, Germany

John Wiley & Sons Australia Ltd, 33 Park Road, Milton, Queensland 4064, Australia

John Wiley & Sons (Asia) Pte Ltd, 2 Clementi Loop #02-01, Jin Xing Distripark, Singapore 129809

John Wiley & Sons Canada Ltd, 22 Worcester Road, Etobicoke, Ontario, Canada M9W 1L1

Wiley also publishes its books in a variety of electronic formats. Some content that appears in print may
not be available in electronic books.

British Library Cataloguing in Publication Data

A catalogue record for this book is available from the British Library

ISBN-13 978-0-470-85506-5 (HB)   978-0-470-85507-2 (PB)
ISBN-10 0-470-85506-1 (HB)   0-470-85507-X (PB)

FSC
Mixed Sources
Product group from well-managed
forests and other controlled sources
Cert no. SGS-COC-2953
www.fsc.org
© 1996 Forest Stewardship Council

Contents

Preface

In an earlier galactic epoch, at least as viewed from the perspective of my present students, I was taught the fundamentals of the science of galaxies by Professor Roger Tayler. While less well known than his classic *The Stars: Their Structure and Evolution*, his undergraduate lecture course on galaxies also generated a text book with a matching title, published in 1978. Though not attempting to match Roger Tayler's concise yet elegant style, the present book *is* intended to be a similar introduction to the study of galaxies at a suitable level – and also short enough – for a middle to late undergraduate course. It may also serve as preparation for the study of more advanced texts and provide an introduction, at least, to topics in extra-galactic astronomy for students wishing to study this fascinating subject at a higher level.

Over the past quarter of a century, our knowledge of galaxies and their evolution has advanced by leaps and bounds. This is reflected in the range of topics which it is now neccessary to introduce in a text book such as this, and excuses (I hope) what may be seen as a reduction in the depth of coverage of some traditional areas. The observations presented, and their interpretation, are intended to be as up to date as is consistent with the time needed for them to become reasonably well-accepted parts of the general picture. The final chapter on galaxy evolution, in particular, attempts to paint a picture which includes recent work (as of the year 2004), but emphasises those points which seem likely to survive the test of time.

A certain amount of background knowledge in mathematics, physics and astronomy – such as might be acquired in any introductory university courses – is assumed. For non-astronomers, an Appendix covers the quaint/bizarre magnitude system which optical astronomers use to measure luminosities and fluxes. Other oddities of units are introduced as required; astronomers are famously unconventional in this regard.

The figures used to illustrate the text are mostly from the professional literature, and are referenced as such. A bibliography of useful texts at a similar or more advanced level is also provided for each chapter. A website associated with this book will be maintained at www.star.bris.ac.uk/sxp/wiley/updates, where new developments can be referenced. In addition, the web is clearly the best place to look nowadays for beautiful images of galaxies. A list of useful sites is included in the bibliography.

Finally, I would like to thank many of my colleagues for reading various drafts of the book, in particular Ian Howarth (for ploughing through it all), Simon Driver, Martin Hardcastle, Nathan Horleston, Bryn Jones and Robin Walker. I must also thank the staff at John Wiley & Sons, especially Rachel Ballard, Andy Slade and Robert Hambrook and all those author who gave the permission to reproduce figures from their papers.

Steven Phillipps
University of Bristol
2004

1 Galaxies in the universe

1.1 Introduction

Galaxies are basically large systems of stars, though we will see as we go along that there is much more to them than that. The Sun and all the stars which we can see at night are part of one such system, the Milky Way Galaxy, or as it is more usually referred to these days, just the Galaxy,[1] with a capital G. As self-contained units, generally well separated from one another, galaxies are often referred to as the building blocks of the universe. In other words, galaxies represent the link between the large-scale cosmological distribution of matter in the universe and the (relatively) small-scale structures around us with which we are more familiar, the Sun and Solar System and our neighbouring stars.

From a cosmological point of view, galaxies also have two key attributes; they are (often) very luminous and there are a vast number of them distributed throughout the universe. Thus galaxies both trace the large-scale structure of the universe and will be visible to us (through large telescopes) even at immense distances. Furthermore, since light travels at a finite speed, we see distant objects as they were in the past, that is, at the time when the photons we see now left the object in question. While the light from the nearest stars takes only a few years to traverse the distance to the Earth, when we look at distant galaxies we will see them as they were millions

---

[1] The name 'galaxy' derives from the Greek word for milk.

The Structure and Evolution of Galaxies   Steven Phillipps
© 2005 John Wiley & Sons, Ltd

or billions of years ago. Not only are we looking out into space, but we are looking back in time as well.

The latter is a particularly pertinent point when we consider that what we see as galaxies today – as described in the subsequent chapters – are the end-points of an evolutionary process which had its ultimate origins in the early universe, immediately after the Big Bang. Since then, as the universe has expanded, galaxies have somehow managed to form and develop, becoming ever more structured and allowing the formation of stars, planets and ourselves. As we shall see, just how this evolution proceeds remains a subject of intense current research. Before that, though, we should start with a look at how our *knowledge* of galaxies has developed.

1.2   A brief history of galaxies

If you have good eyesight you can see up to about 2000 stars at any one time. If you are lucky enough to be able to view the night sky from a dark site well away from artificial lights – an unlikely circumstance for most modern town dwellers, unfortunately – at the right time of year you can also see crossing the sky a pale cloudy band. This is the Milky Way. When Galileo first turned his new telescope to the heavens in 1610, he immediately saw that the Milky Way was in fact made up of vast numbers of faint stars.

The reason why all these stars appear as a band around the sky was correctly deduced by Thomas Wright in about 1750. The stars around the Sun are distributed in a flattened system – somewhat like the shape of a discus in athletics – so when we look in directions in the plane of this disc we see many stars along the same line of sight, but when we look out of the plane on either side we see relatively few stars (Figure 1.1). Towards the end of the 18th century this observation was placed on a quantitative footing by William Herschel, following his monumental efforts in 'star gauging' (i.e. counting) in different directions on the sky.

If you are observing from the Southern Hemisphere, you can also see what look like separate broken-off patches of 'cloud' away from the Milky Way itself. These are the Large and Small Magellanic Clouds, named after the explorer Ferdinand Magellan as they were first reported in Europe in 1522 following Magellan's round-the-world voyage (though of course they were already known to people such as the Australian aborigines, who included them in their dreamtime myths). The Magellanic Clouds are in fact separate stellar systems, now known to be smaller satellites or companion galaxies of our own Galaxy.

In the Northern Hemisphere, another even smaller cloudy patch is just visible to the naked eye (and easily seen with binoculars) in the constellation of Andromeda. This is the famous Andromeda Nebula ('nebula' being the Latin word for cloud). Surprisingly, this was not recorded convincingly by European astronomers until after the invention of the telescope, although it had been included on Arabic star maps in the 10th century. The Andromeda Nebula turns out to be the nearest external giant galaxy comparable to (in fact slightly larger than) our own. However, the fact that these fuzzy patches – and many other fainter ones – are truly external star systems in their own right was only conclusively proven in the 1920s.

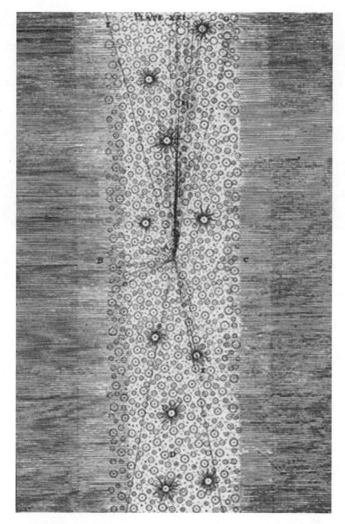

Figure 1.1 Illustration from Thomas Wright of Durham's 'An Original Theory of the Universe', 1750. (Courtesy: The Norton Anthology of English Literature.)

Starting with Messier's famous catalogue (of objects to avoid when hunting comets!), published in 1784 (and from which names such as M31 for the Andromeda Nebula arise), a large number of nebulae were discovered during the 18th and, especially, 19th centuries, many of them by William Herschel and his son John. Indeed, by 1908, Dreyer's New General Catalogue (NGC) and Index Catalogue (IC) contained entries for around 13 000 nebulae and star clusters. However, astronomers were divided as to whether they were relatively small systems within what we now call our Galaxy, or large star systems outside our own. The latter idea, originating again with Thomas Wright and the philosopher Immanuel Kant, was known as the 'island universe hypothesis'.

Figure 1.2   The whole 360° view of our Galaxy from the inside. The Galactic Centre is in the centre of the plot, the anti-centre direction is at the edges. Also visible in the southern Galactic hemisphere (below the main central plane, the Milky Way) are the Magellanic Clouds. (Courtesy: Lund Observatory.)

Between 1845 and 1850, Lord Rosse had discovered spiral structure in some nebulae, most famously M51, otherwise known as the Whirlpool. This was taken, by some astronomers at least, to imply small, fairly nearby systems, since a then popular theory of the formation of planetary systems around stars had them start from just such beginnings. In the 1860s, with the dawn of spectroscopy, William Huggins found that the spectra of some nebulae implied that they were gas clouds (and hence 'nebulae' in

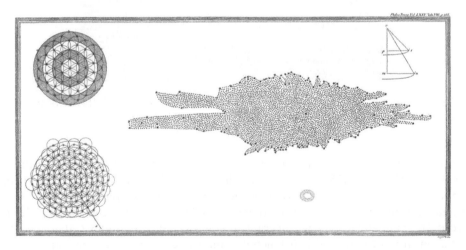

Figure 1.3   Herschel's sketch of the Galaxy, representing the numbers of stars seen in different directions on the sky, from 'The Construction of the Heavens', Phil. Trans. Roy. Soc., 1785. The twin-tailed appearance at the left of the plot – now known to be the direction towards the Galactic Centre – is due to dust obscuration of the stars close to the central plane.

the modern usage of the term), and, to add to the problems for the island universe side, Proctor demonstrated in 1869 that the distribution of nebulae on the sky was such as to avoid the Milky Way disc, surely evidence for a physical connection.

In 1885 a nova (i.e. 'new star'), S And, was seen in the Andromeda Nebula. This had a luminosity around one-tenth of that of the whole nebula. Other novae had been seen in our Galaxy and their luminosities were known, at least roughly. If the nova in Andromeda could be assumed to be similar, then its apparent brightness certainly implied that it must be well within the Galaxy. Thus Scheiner's discovery that the nebula actually had a spectrum like a star (or large number of stars) was not seen as convincing evidence for an external star system. Note that at this time, the Galaxy was thought to be perhaps 5–10 kpc in diameter; so if the same was true for M31 then, using a simple argument still much in use today, its angular size θ of about 1 degree (or 0.02 radians) and physical size d implied a distance

$$D = \frac{d}{\theta} \simeq 250 - 500 \, \text{kpc} \qquad (1.2.1)$$

In 1916 the discussion between the 'island universe' and the 'metagalaxy' factions was sidetracked somewhat, when van Maenen reported that he had measured internal motions within the spiral nebula M101. If the spiral nebulae really were at very large distances, it should have been impossible for anything to move fast enough to generate a measurable change in position on the sky over a few years. However, this remained a controversial observation, and ultimately proved to be an unexplained error. Evidence by now had begun to sway towards the island universe theory with, for instance, the discovery of what appeared to be many more novae in spiral nebulae. These appeared thousands of times fainter than S And and compatible with distances of order Mpc for the nebulae if they, and not S And, were the equivalents of Galactic novae.

The year 1920 saw the so-called 'Great Debate' at the American Society for the Advancement of Science, between Heber Curtis for the island universe side and Harlow Shapley for the single metagalaxy. Curtis countered the usual arguments against external systems by positing that S And was, in fact, a 'supernova', much more luminous than conventional novae, and that nebulae were concentrated towards the Galactic polar regions not because of a physical association, but because dust between the stars obscured those which would otherwise be seen through the plane of the disc-like Galaxy. Most astronomers probably favoured the idea of external galaxies by this time, but definite proof was still lacking. Clearly what was needed was a direct measurement of the distances to the nebulae.

1.3   Distance measurements

We should backtrack slightly at this point, to consider how the scale of the Galaxy had been established. Even as far back as Tycho Brahe in the 16th century, astronomers had appreciated that it should, in principle, be possible to measure the distances to the stars by using measurements of trigonometric parallax, that is the change in the apparent direction of the star when viewed from two different positions. This method was used successfully to measure the distance to Mars, for instance, just by using observers at two well-separated points on the Earth. To obtain a larger baseline, and hence be able to measure out to greater distances, we can utilise the movement of the Earth around the Sun; by making observations six months apart we can obtain a baseline of 2 AU. (1 Astronomical Unit is, by definition, the mean distance of the Earth from the Sun, 149.6 million kilometres). A measurement with a baseline of 1 AU is called a star's 'annual parallax' (Figure 1.4).

However, due to the large distances to the stars, no successful parallax measurement was obtained until 1838. The man finally to succeed was Friedrich Bessel, who obtained a parallax of just 0.29″ for the star 61 Cygni.[2] Bessel had chosen this star for his attempt at the parallax measurement because he knew that it had a large 'proper motion', that is it appeared to move systematically across the sky (relative to the positions of neighbouring stars) at a faster rate than almost any other known star. Assuming similar physical velocities for all stars, the closer ones will clearly appear to move faster, though 'faster' is only a relative term here, as even 61 Cygni moves only about 5″ per year (so would take 350 years to cross a distance equal to the diameter of the Moon).

With the definition of the parsec as the distance at which a star would have an annual parallax of exactly 1 arcsecond (i.e. 3.086×10^{16} m), Bessel's measurement implied a distance of about 3.5 pc for 61 Cygni. Subsequent observations have confirmed Bessel's presumption that it should be one of the nearest stars to us, though Proxima Centauri, part of the α Centauri multiple star system, is actually the closest at 1.3 pc.

In fact, William Herschel (and Christiaan Huygens, a century earlier still) had preempted the actual measurement of a stellar distance by making another simple argument which is still regularly used as a substitute for real distance measurements. The scale of the Solar System, and hence the distance to the Sun, was already known by Herschel's day. If we assume that the other stars that we see are (at least on

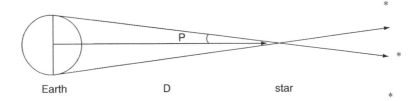

Figure 1.4   Trigonometric parallax. As the Earth orbits the Sun, the observer will view a given nearby star from positions separated by up to 2 AU. Half the apparent shift of the star's position on the sky relative to more distant objects (the same as the angle subtended by the Earth's orbital radius as seen from the star) is called its annual parallax p.

---

[2] Strictly speaking, the first person to *measure* a parallax was Thomas Henderson in 1832, but his result was not published until after Bessel's.

average) of similar luminosity to the Sun, we can estimate how much further away they must be in order to look as faint as they do.

Couched in modern terms, given the distance D to the Sun and the measured flux of sunlight at the Earth ($F_\odot = 1.37\,\mathrm{kW\,m^{-2}}$), the luminosity of the Sun (i.e. its power output, $L_\odot$) can readily be deduced by using the simple inverse square law:

$$F_\odot = \frac{L_\odot}{4\pi D^2} \tag{1.3.1}$$

This gives $L_\odot = 3.86 \times 10^{26}\,\mathrm{W}$.

From the fluxes of the brightest stars and the assumption $L \simeq L_\odot$, another application of the inverse square law shows that the brightest stars should be of order 2×10^5 times further from us than the Sun, that is about 1 pc away, a rather good estimate. Indeed, Herschel went a step further. If stars were typically 1 pc apart, then from the volume density of stars that this implies, and the total number of stars which he could see, he could deduce that the overall size of the Galaxy must be about 800 (in the plane) by 150 pc (perpendicular to the plane). By 1900, more sophisticated methods along the same lines led to the 'Kapteyn Universe', that is a star system 7000 by 1300 pc.

A key breakthrough in the distance measurement problem came in 1908 as a result of Henrieta Leavitt's study of Cepheid variable stars in the Magellanic Clouds. Cepheids have a very characteristic variation in brightness with time, and Leavitt discovered that there was a relationship between the period of the variations and the average brightness of the star. Since all the Cepheids in one of the Clouds could be assumed to be at essentially the same distance, this clearly translated into a relationship between their periods (P) and (relative) luminosities (L).

To calibrate this relationship, one needs to know the absolute luminosity of one or more nearby Cepheids, and to determine that one needs to know their distances. Cepheids are relatively rare, and none was near enough so that a trigonometric parallax could be obtained. Harlow Shapley therefore used the method of 'statistical parallax', which is not really a parallax at all, but utilises the velocities of stars (section 1.6).[3]

Once it was calibrated, Shapley was able to utilise the Cepheid P–L relation to determine the luminosity of any Cepheid with an observed period, and then by using its apparent brightness (and the inverse square law) he could deduce its distance. In this way, in 1915, Shapley obtained distances to a number of globular clusters (GCs) (large agglomerations of perhaps a million stars, in a densely packed, almost spherical system a few pc across) in which he could find Cepheids, showing that they ranged up to about 50 kpc.

Shapley also used GCs to demonstrate that the Sun was not at the centre of the Galaxy. Though globulars are distributed all around the sky, there is a preponderance in one direction. If we make the reasonable assumption that they should

---

[3] Strangely, in his 1943 book *Galaxies*, Shapley implies that he did determine the absolute luminosities via ordinary trigonometric parallaxes.

be symmetrically placed with respect to the centre of the Galaxy, then this observation can easily be explained if we are viewing them from an off-centre position. Since Shapley now also had a reasonable idea of the distance to many globulars, either directly from Cepheids or from other types of star calibrated with respect to them, he was able to estimate that the Sun must be around 10 kpc from the Galactic Centre.

Ironically, the eventual resolution to the question of the distances to the nebulae used Shapley's Cepheid method to find in favour of his opponents. In 1923, Edwin Hubble used the new 100″ telescope at Mount Wilson to discover Cepheids in M31.[4] With the known *P–L* relation, he could then show that M31 must be at least 300 kpc away, clearly in line with it having a size comparable to our own Galaxy. Further discoveries of Cepheids in other spiral and irregular nebulae soon followed, and the case for external galaxies was made when Hubble published his results in 1925.

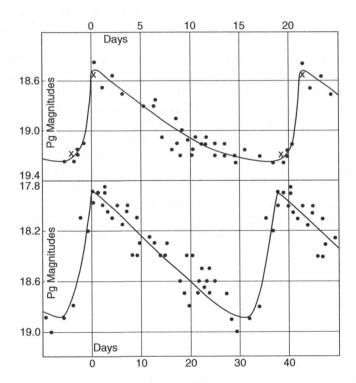

Figure 1.5 Hubble's first published Cepheid observations in an external galaxy, NGC 6822 (Reproduced with permission from Hubble, 1925).

---

[4] According to Milton Humason, originally an assistant to Shapley but who later worked with Hubble, he (Humason) had discovered candidate Cepheids in M31 prior to this, but Shapley, not believing this to be possible, 'calmly took out his handkerchief, turned the [photographic] plate over and wiped it clean' (of Humason's identifying marks).

1.4   Redshifts, distances and dynamics

In retrospect, the decision could have come even earlier. As mentioned in the previous section, spectroscopy had been applied to astronomy (creating the new science of astrophysics) from the 1860s. Apart from anything else, this soon provided a new tool for making measurements of stars' velocities. The Doppler Effect on sound waves – the squashing up and therefore shortening of wavelength of sound waves from an approaching source (and the reverse for a receding source) – was already well known, and an analogous result holds for light waves from a moving source. For a source moving away from an observer at speed ν, for example, we can simply assume that each successive wave crest, emitted at time Δt_{em} apart, will have an extra distance $\nu \Delta t_{em}$ to travel and therefore takes an extra time $(\nu/c)\Delta t_{em}$ to reach the observer. Thus the observer sees the wave crests separated by time intervals

$$\Delta t_{obs} = (1 + \nu/c)\Delta t_{em} \qquad (1.4.1)$$

and therefore sees the wavelength of the light stretched by this same factor $1 + \nu/c$.

If we make a general definition of redshift z via the relation

$$1 + z = \frac{\lambda_{obs}}{\lambda_{em}} \qquad (1.4.2)$$

where λ_{obs} and λ_{em} are the observed and emitted wavelengths, then to first order in velocity (as we have neglected relativistic effects in this simple derivation), we will have

$$z = \frac{\Delta\lambda}{\lambda_{em}} = \frac{\nu}{c} \qquad (1.4.3)$$

where $\Delta\lambda = \lambda_{obs} - \lambda_{em}$ is the change in wavelength seen for lines in the spectra of a source moving at speed ν.

In this way, Huggins and others determined stellar radial velocities to be typically tens of $km\,s^{-1}$. In 1912, after a monumental observing campaign at Lowell Observatory (he re-exposed on the same photographic plate every clear night for a month!), V.M. Slipher managed to obtain a spectrum of M31 with sufficient detail to measure its redshift. In fact, what he actually measured was a blue shift – the lines being shifted to shorter wavelengths due to a velocity of around 300 $km\,s^{-1}$ *towards* us. Over the next few years, though, further observations showed that the large majority of spiral nebulae observed were moving away from us, some of them at velocities up to 2000 $km\,s^{-1}$. These were much greater than the velocities seen for stars, and in fact are much too large for the nebulae to be Galactic objects of any sort: their velocities far exceed the escape velocity from the Galaxy. Slipher, however, remained in the metagalaxy camp until this point about the escape velocity was made to him some years later.

A further use of redshifts, also pioneered in the very early days of nebular spectroscopy, comes in the study of the internal dynamics of the objects observed. For instance, in a rotating disc galaxy seen edge on, one side will be approaching the observer and the other side receding. Thus one side will show a blueshift and the other a redshift, enabling the rotation speed to be determined. Interestingly, this also provided another early, but generally neglected, pointer to spiral nebulae as external systems. As pointed out by Öpik in 1922, the observed rotation velocity for M31 can be used to estimate its distance. If we assume that the stars at the edge of its visible disc are moving in circular orbits at velocity V_c under the gravitational attraction of the bulk of the galaxy's mass, M (just as the planets rotate about the Sun), then the gravitational force must match the centripetal acceleration required for circular motion. Thus

$$\frac{GM}{(\theta D)^2} = \frac{V_c^2}{\theta D}$$

(1.4.4)

where we have written the physical radius as the measured angular radius θ times the unknown distance D.

To get at the mass M, Öpik noted that on average in our Galaxy, to produce the same power as generated by the Sun (one solar luminosity or $1L_\odot$) we require about 3 solar masses of stars ($3M_\odot$). (This is because the majority of stars are smaller than the Sun and low mass stars are proportionately less efficient in the power generation stakes). This is conventionally written as a mass-to-light ratio $M/L \simeq 3$, where the normalisation by solar values is usually taken as read. Now Öpik could also determine the luminosity of M31 in terms of its apparent flux, F, and the distance D,

$$L = 4\pi D^2 F$$

(1.4.5)

so combining all these results

$$D = \frac{V_c^2 \theta}{4\pi GF(M/L)}$$

(1.4.6)

Using the observed values for the terms on the right-hand side of this equation, he was thus able to deduce a distance, arriving at an estimate of about 450 kpc. Though used here to illustrate a point regarding the distances to nebulae, such observations of the rotation of spiral galaxies and their stellar contents have remained at the forefront of work in more recent times, such as the deduction of the presence of dark matter, as we shall see in Chapter 4.

1.5   Expansion of the universe

Returning to the development of extra-galactic astronomy, as it had now become, we come to Hubble's second great epochal discovery. As we have seen, Slipher, and subsequently others such as Milton Humason at Mount Wilson, found that almost all other galaxies were moving away from us. By one of the remarkable juxtapositions of theory and observation that seem to characterise astrophysics and cosmology, Albert Einstein was at the same time formulating his General Theory of Relativity (GR), and by the early 1920s Arthur Eddington and others had already considered the possibility that the 'recession of the nebulae' was a General Relativistic effect.

Hubble – who always professed himself sceptical of theory – was the one to make the big leap forward, though. Continuing his use of Cepheids in nearby galaxies, and then extending the work by using as a so-called secondary distance indicator the apparent brightness of the most luminous stars in a given galaxy (i.e. by assuming that the most luminous stars are always physically the same, regardless of the galaxy they are in), he was able to estimate distances to 18 galaxies with redshifts up to $1000 \, \mathrm{km \, s^{-1}}$. Despite the large scatter in the points, in a classic paper in 1929 Hubble presented a 'roughly linear' relationship between recession velocity cz (equation 1.4.3) and distance D, a result extended to greater distances in a paper with Humason two years later.

A few years earlier, Alexandr Friedmann and the Abbé George Lemaître had (independently) shown that there were solutions of Einstein's equations of GR, as applied to the universe as a whole, which allowed uniform expansion (or contraction). Hubble's observational result was immediately associated with such a general expansion,[5] as it requires redshift proportional to distance, and thus marked the beginning of modern observational cosmology.

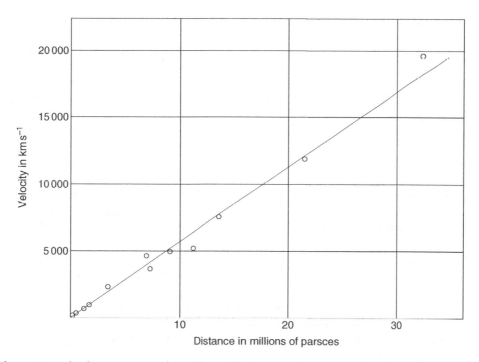

Figure 1.6 The first extensive plot of the redshift versus distance relation (Reproduced with permission from Hubble and Humason, 1931).

---

[5] Indeed, Hubble's paper was predated by papers by Lemaître and another relativist H.P. Robertson, which suggested that Slipher's and Hubble's results on redshifts and distances might be interpreted in terms of expansion.

Aside from the cosmological importance, the key application of Hubble's law is that it provides a means of assigning a distance to any galaxy for which a redshift can be obtained, regardless of any actual distance measurement via Cepheids or any of the other methods discussed below. Unsurprisingly, the constant of proportionality H_0 in the law

$$cz = H_0 D \qquad (1.5.1)$$

is now known as Hubble's constant. Because of its origin, H_0 is conventionally written with units of $\mathrm{km\,s^{-1}\,Mpc^{-1}}$. However, we can also see that, as a ratio of a velocity and a distance, it must have the dimensions of $(\mathrm{time})^{-1}$. Indeed, to a first approximation (of no change in the expansion rate), $1/H_0$ – the Hubble time – must represent the age of the universe, the time since the Big Bang, since at speed cz it will take a time $1/H_0$ to travel the separation D. Note, too, that the quantity c/H_0, the distance that light can travel in a Hubble time, will similarly give a characteristic scale for the size of the possibly observable universe – the Hubble radius.

1.6   Hubble's constant and the distance scale

As we have seen, Hubble's original demonstration of the law which bears his name, derived from a handful of distance estimates, based on either Cepheid variables or the brightest stars in galaxies. Even inside our own Galaxy, distance determinations rapidly become insecure once we pass beyond the realm where trigonometric parallaxes are possible. Until the 1990s this limit was at tens of parsecs because of the practical limitation on measuring angles much less than 0.1 arcsecond (the angle subtended by a 200-m diameter crater on the Moon). Since then, the precision allowed by satellite-based observations, particular from Hipparcos, has enabled us to extend these measurements to stars at distances of up to 1 kpc. In the absence of precise measurements, astronomers were particularly inventive in their attempts to measure distances, though, and concocted a whole series of alternatives, useful for different distant ranges, and between them forming the so-called 'cosmic distance ladder'.

Some of these involve observing the motions of groups of stars, for instance the 'moving cluster' method, traditionally used to determine the distance to nearby star clusters like the Hyades. Here one assumes that all the stars are moving around the Galaxy together, so can be assumed to have parallel (3-D) velocity vectors. Just as physically parallel railway lines appear to converge in the distance, so the motions of the stars will appear to be towards some 'convergent point' on the sky. By combining measurements of the angular distance of a star from this convergent point (θ) with its radial velocity ν_{r} and proper motion μ (Figure 1.7), it is straightforward to deduce that its distance (in parsecs) should be

$$D = \frac{\nu_{\mathrm{r}} \tan \theta}{4.74 \, \mu} \qquad (1.6.1)$$

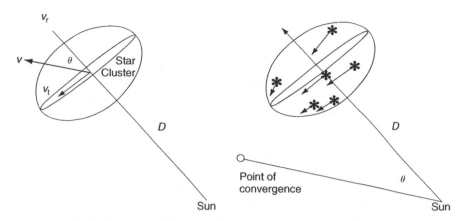

Figure 1.7   The Moving Cluster method. All the stars in the cluster are assumed to be moving, with velocity v, along parallel tracks which appear to converge at the Convergent Point. (Figure courtesy of Diana Worrall.)

since the tangential component of the velocity (i.e. across the sky)

$$\nu_t = \nu_r \tan\theta \qquad (1.6.2)$$

is also measured by the proper motion (in arcsecond per year) multiplied by the distance (which, it is easy to check, gives the motion in AU/year). The factor 4.74 comes from the translation between $D\mu$ in AU/year and ν_r in $\mathrm{km\,s^{-1}}$.

The statistical parallax method alluded to earlier is somewhat related to this in that we use the statistical properties of a whole set of stars. In particular, we assume that, once the overall orbital motions are allowed for, then on average the velocities of the stars should be the same across the sky (where they are determined by the measured proper motions and the unknown distances) and in the radial direction (where they are measured directly from Doppler shifts). Thus we can deduce the mean distance to the sample stars.

Once we have the distances to some clusters, we can utilise the global characteristics of stellar populations. Around 1910, Ejnar Hertzsprung and H.N. Russell both realised that if they plotted the temperatures of stars, as obtained from their colours, against their intrinsic luminosities – usually expressed as absolute magnitudes[6] – then only certain characteristic regions of the plot (now known as the Hertzsprung–Russell, or H–R, diagram) were actually populated. Indeed the majority of stars occupied a swathe from bright and blue (hot) to faint and red (cool). This is the stellar 'main sequence', on which stars spend most of their lives.

Now, if we have a relationship between the colour (or spectrum) and the luminosity of stars, calibrated by measurements on a cluster at known distance, then we can clearly use this as a distance indicator. This is known as 'spectroscopic parallax',

---

[6] Recall that optical astronomers habitually use the logarithmic magnitude scale rather than actual luminosities. Readers unfamiliar with magnitudes should refer to the Appendix or any introductory astronomy text book.

though again no real parallaxes are involved. In the absence of any absorption of the light by intervening interstellar material, the colour will be independent of the distance, but the apparent luminosity will follow the usual inverse square relationship. Thus we can determine the distances to many more stars, just by measuring their colours and apparent magnitudes. This is especially profitable if we look at further star clusters; as we will essentially see our standard main sequence just shifted in magnitude by the 'distance modulus' to the cluster, the difference between the apparent magnitudes m and absolute magnitudes M is given by

$$m - M = 5 \log_{10}(D/10) \tag{1.6.3}$$

for D in pc.[7] This method of distance determination is known as 'main-sequence fitting'. If the cluster happens to contain a particularly 'useful' sort of star (perhaps some characteristic type of variable star such as an RR Lyrae) then we will be able to calibrate *its* absolute brightness and hence use that as another 'standard candle' (i.e. source of known brightness) to determine yet more distances. As well as the main sequence, the H–R diagram of a typical cluster will also contain a 'giant branch', composed of stars which have evolved off the main sequence when the hydrogen fuel

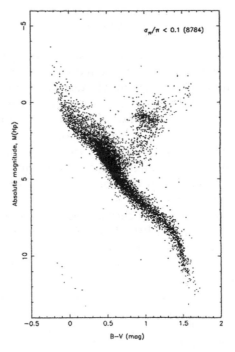

Figure 1.8   The H–R diagram for nearby stars, as determined from measurements with the Hipparcos satellite, showing the main sequence and the giant branch (towards the upper right). (Reproduced with permission from Perryman *et al.*, 1995).

---

[7] Recall that the absolute magnitude is defined as the magnitude an object would have if placed exactly 10 pc away.

in their central parts has been exhausted. These stars expand to huge sizes, so although quite cool (hence 'red giants'), they have large luminosities. It turns out that the stars at the tip of the red giant branch (TRGB) have quite characteristic luminosities, so this TRGB provides yet another useful distance indicator which can be used for any star cluster, or indeed external galaxy in which individual stars can be resolved. This is essentially a more modern version of Hubble's assumption that the brightest stars in any galaxy should be of the same absolute brightness.

Beyond the distance where individual stars could be seen (because they appeared too close together to be separated in the presence of the blurring – called 'seeing' – due to the Earth's atmosphere), Hubble and his successors had to rely on even less direct means, such as the size of the largest HII regions (luminous regions of ionised hydrogen) in a galaxy, or the appearance of the whole galaxy. Indeed, to reach the greatest distances we can try to repeat Hubble's trick with the stars, and assume that the brightest galaxy in any cluster of galaxies always has (more or less) the same absolute magnitude. Recently there has been considerable success in using super-novae as standard candles. We will return to some of these methods in subsequent chapters, but many have now been superseded.

Unsurprisingly perhaps, the range (and somewhat dubious precision) of the distance indicators employed led to uncertainty – indeed controversy – over the value of Hubble's constant, the ratio of recession velocity to distance. Hubble himself originally estimated that its value was $550 \, km \, s^{-1} \, Mpc^{-1}$, but in the 1950s it was shown that this was a considerable overestimate (distances had been underestimated) because of (a) confusion between true Cepheids and the very similar (but different luminosity) W Virginis stars found in GCs and (b) some of Hubble's brightest 'stars' turning out to be star clusters or HII regions.

A further uncertainty is introduced into distance measurements by the fact that spiral galaxies such as our own not only contain stars, but also possess an 'interstellar medium' (ISM). The ISM is a mixture of gas and dust which acts like a 'fog' between the stars. This will obviously make the stars look dimmer than they should be on the basis of their distance alone, and therefore complicates all our distance estimates based on apparent brightnesses and the inverse square law.

Subsequent to the reduction in Hubble's value of H_0, measurements (and personalities) polarised into two camps. One, mainly associated with Gerard de Vaucouleurs, maintained that the value of H_0 was close to $100 \, km \, s^{-1} \, Mpc^{-1}$, the other, led by Allan Sandage, that it was instead close to $50 \, km \, s^{-1} \, Mpc^{-1}$. Notice that this was equivalent to a disagreement by a factor of two between the so-called 'short'- and 'long'-distance scales. A large fraction of this difference was already present[8] in the distances to their local calibrating galaxies.

With the advent of the Hubble Space Telescope (HST) in the 1990s, however, it became possible to detect Cepheids at much greater distances. This then allowed many more galaxy distances to be determined reasonably directly without getting into the complexities of the distance ladder. In addition, it allowed the determination

---

[8] Indeed, reputedly, the only thing that the two camps agreed on was that the real answer would not turn out to be halfway between their respective values. Nonetheless, it did!

of distances of galaxies sufficiently far away that their 'peculiar velocities' (the local motions they have due to the gravitational effects of their neighbours) are small compared to their 'Hubble flow' velocities due to the overall expansion of the universe. This removes one source of uncertainty in the relation between distance and expansion rate. Another problem can be alleviated if we work in the near infra-red, since any absorption of light by intervening interstellar dust (either in our Galaxy or in the target galaxy) is thereby reduced. In this case, the *P–L* relation for Cepheids can be written in terms of *I*-band magnitudes as

$$M_I = -1.3 - 3.5 \log P \tag{1.6.4}$$

for *P* in days. Thus a Cepheid with a period of 100 days, for example, will have an *I*-band magnitude of −8.3 and be easily visible with HST (which can routinely reach magnitude limits of $m_I = 25$) out to a distance modulus $m - M \simeq 33$ corresponding to a distance of 40 Mpc, comfortably beyond the Virgo Cluster, the nearest really large grouping of galaxies. The HST 'key project' on the distance scale, carried out by Wendy Freedman and colleagues, and other recent work have now settled on a value close to

$$H_0 = 70 \pm 2 \, \text{km s}^{-1} \, \text{Mpc}^{-1} \tag{1.6.5}$$

This then leads to a Hubble time

$$\frac{1}{H_0} = \frac{3.086 \times 10^{22}}{70} = 4.41 \times 10^{17} \, \text{s} = 13.95 \, \text{Gyr} \tag{1.6.6}$$

(using the number of km in an Mpc and seconds in a year) and a Hubble radius

$$c/H_0 \simeq 4300 \, \text{Mpc} \tag{1.6.7}$$

1.7   The observable universe

Given these figures, and before moving on to the galaxy population itself, we should perhaps end this introduction by considering just how far out in the universe, and thereby back in time, we are able to observe galaxies. It is often more convenient to do this in terms of the redshift defined earlier rather than distance, since the former is directly observable and the latter is not. On local scales, we have looked on the redshift as a recession 'velocity'. In fact, this is not strictly correct, as the cosmic expansion increases the space between the galaxies, without them actually moving; the whole fabric of the universe expands, as in the usual analogy of blowing up a balloon with dots for galaxies drawn on it. Thus the redshift can be arbitrarily large, without conflicting with the fact that nothing can travel faster than the speed of light. We will look at these cosmological considerations in detail in the final chapter, but for now we will just use redshift as a convenient flag for distance.

As a normalisation, we can note that the redshift of the Virgo Cluster, around 20 Mpc away, is about $z = 0.005$. However, much more distant galaxies are seen even at quite bright apparent magnitudes and by the 1950s galaxies were already known at redshifts beyond $z = 0.1$. Then in 1960, the radio source 3C295[9] was identified with a galaxy in a cluster and found to have a redshift of 0.46. Although 'Minkowski's Cluster' remained the furthest known for many years, the overall distance record was soon surpassed by an entirely new type of object, when Maarten Schmidt identified another class of radio source, quasars. These are now known to be extremely luminous 'active' nuclei of galaxies. In 1965, 3C9 was the first object found with a redshift of 2 and by 1973 the most distant known object was another quasar at $z = 3.73$. The first redshift-4 quasar followed in 1987 but, by this time, observations of other sorts of galaxy were catching up again, with radio galaxies found out to $z \simeq 1.8$. Indeed, in 1988, an otherwise anonymous 24th magnitude normal (non-active) galaxy was found to have $z = 3.345$, and after ultra-deep imaging observations were made using the HST (in the 'Hubble Deep Field' (HDF)), one of the galaxies therein broke the $z = 5$ barrier. Finally (as of the time of writing), the last couple of years has seen the first $z \simeq 6$ quasars and galaxies identified. Current cosmological models imply that we are seeing these object as they were around 13 billion years ago, or only about 1 billion years after the Big Bang. Clearly, at some point not much beyond this, we will reach a limit to our observations of galaxies, as we will be looking back to a time before galaxies had formed.

Though not related to galaxies in the usual sense, there is one other observable component of the universe that we should mention. This is the 'cosmic microwave background' or CMB. This is often referred to as 'relic' radiation from the Big Bang, as it originates from the epoch around 200 000 years after the Big Bang when the universe – cooling as it expanded – was first able to form neutral atoms. In the process the universe became transparent to radiation; previous to this epoch – called 'recombination' – the high density of free electrons in the universal plasma led to the immediate scattering of any photons, making the universe opaque. The photons present at recombination have since been redshifted down in frequency (and hence energy) to their present temperature of 2.7 K and hence are prominent in the microwave region of the spectrum.

[9] An object in the 3rd Cambridge catalogue of radio sources.

2 A galaxy menagerie

2.1 Morphological types

The galaxies which Hubble, and before him the likes of Lord Rosse, studied were often spiral in form, like M31 and M81. However, not all galaxies are spirals; indeed our nearest neighbours, the Magellanic Clouds, present little, if any, organised structure and are prototypes of the general class of irregular galaxies. Somewhat further afield (but still bright enough to be in Messier's catalogue), we meet the first examples of the other main variety, or 'morphological type', of large galaxy. These are the ellipticals, apparently regular in shape and free of substructure, such as the famous M87 in the constellation Virgo.

In most cases, the appearance of elliptical galaxies depends just on the population of stars they contain. Early observers already knew that ellipticals are 'red', in the sense of emitting more light at longer wavelengths, so should contain mostly red stars. Spirals, on the other hand, contain gas and dust between the stars, and the spiral patterns we see are actually delineated by both dark dust lanes and bright regions where new stars are forming from this material. As a population of young stars will contain bright blue ones, spirals are also bluer in colour than ellipticals.

The first real attempt at classification was made by Reynolds in the early 1900s, even before galaxies' external nature was certain. However, it was Hubble (again!) who came up with the system which still serves as the starting point for all more complex classifications. This was the famous 'tuning fork' diagram shown in Figure 2.1. The handle of the fork comprises the elliptical galaxies, running from circular (as projected on the sky) E0s to more and more flattened E1s–E7s. Here the numeral (0–7)

The Structure and Evolution of Galaxies   Steven Phillipps
© 2005 John Wiley & Sons, Ltd

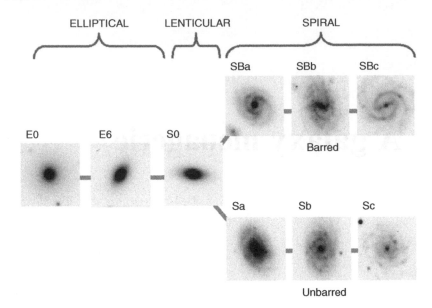

Figure 2.1 Hubble's 'tuning fork' diagram for galaxy classification, with elliptical galaxies in the 'handle' and spirals in the 'prongs'. (Figure courtesy of Simon Driver; images from the Isaac Newton Telescope.)

represents the elongation of the galaxy's image via the quantity $10(1 - b/a)$, where b/a is the ratio of the short (minor) to long (major) axis lengths. Thus an image with an axis ratio of 0.7:1 will correspond to an E3, for example.

The prongs of the fork represent spiral galaxies. While, as one might presume, elliptical galaxies are ellipsoidal in their 3-D shape, spiral galaxies, like our own, are basically flat discs. Thus they can present very elongated images if seen sufficiently edge-on, though they do often also have a central spheroidal component known as the 'bulge'. Viewed face-on, spirals present many different detailed forms. First, the characteristic spiral pattern, produced by the presence of bright spiral arms, can be more or less tightly wound. Sa galaxies have the most tightly wound arms, while the pattern in Sb, Sc and Sd galaxies becomes progressively more open (i.e., the 'pitch angle' of the spiral becomes larger). In step with this change in the spiral arm winding, spirals show a decrease in the importance of their spheroidal components, that is Sas have large bulges while Scs have small bulges. Indeed, bulges are almost non-existent in Sd galaxies (which are actually an addition to Hubble's original classes).

At the end of the sequence, Hubble placed the Irregular systems, which are again basically 'flat' but with chaotic patterns of bright regions. An intermediate class, the Sms, with rudimentary or fragmentary arm-like structures was added later on, and Irregulars are nowadays often referred to as Im galaxies, the 'm' in each case standing for Magellanic, after the proto-typical Clouds.

The reason for there being two prongs to the fork is that spirals can also be separated into two parallel sequences depending on whether the spiral arms emanate from the central bulge or from the ends of a further component, a central 'bar' (reminiscent of a bar magnet placed across the bulge). The barred galaxies then provide the types SBa, SBb, etc. Roughly half of all spirals appear to be barred.

Subsequent to the original classification, Hubble added another category, the S0 galaxies. These occupy the position where the prongs join the handle. They possess large bulges and disc components, but show no sign of any spiral pattern, so are also known as lenticular galaxies on account of their lens-like shape. They can be viewed as intermediate in form between the ellipticals and the spirals. Elliptical and S0 galaxies are often referred to jointly as 'early type' galaxies, and Sas as early type spirals, while Scs etc. are late type spirals. This is because Hubble conjectured that they might form an evolutionary sequence. This idea soon fell into disfavour, and for several decades text books decried the notion. More recently, the idea of transmutation between different morphological types has been revived, but now mostly in the reverse sense to that proposed by Hubble!

In the 1950s and 1960s more elaborate classification schemes for spirals were devised, primarily by Gerard de Vaucouleurs, Allan Sandage and Sidney van den Bergh. Both de Vaucouleurs and Sandage built directly upon Hubble's system, introducing the Sds and Sms and filling in intermediate types such as Sab or Sbc, and the intermediate family of weakly barred systems, SABa and so on. (In this system the original unbarred galaxies are SAa, etc.) Also introduced were the further descriptors 'r' for ringed, 'rs' for weak rings and 's' for no ring, where a ring is defined as a more or less circular structure towards the centre of a galaxy (but outside the bulge) from which the arms appear to start. Thus a galaxy can have a detailed classification like SB(r)ab. de Vaucouleurs also introduced a numerical sequence to represent the main morphological types. These run from $T = -5$ up to -1 for ellipticals through to S0s, 0 for S0/a galaxies, 1 for Sa galaxies and so on up to 5 for Sc and 9 for Sm, that is one numerical class for each Hubble sub-class. Class 10 is used for irregulars.

van den Bergh's classification, on the other hand, was an attempt to indicate the intrinsic luminosity of a galaxy. He found that (with a fair amount of scatter), the luminosity of a spiral galaxy correlated with the structure of its arms. More luminous spirals had better developed, well-defined continuous arms, while low luminosity ones had weak, patchy arms. The former are sometimes called 'grand design' spirals, the latter 'flocculent' spirals. A luminous Sc galaxy with very clearly defined arms would then be an ScI, while a less bright one with indistinct arms might be an ScIII, the Roman numerals denoting the 'luminosity class'.

2.2   Luminosities and sizes

This last point leads us nicely on to the question of the luminosities of galaxies in general. Observationally, it is clear that if we know the distance to a galaxy (or can infer it from its redshift) and can measure its apparent brightness, then we can use the inverse square law to deduce its intrinsic luminosity (power output). When discussing galaxies (at least at optical wavelengths) astronomers seldom work in conventional 'physics' units such as Watts, usually preferring Solar luminosities, $L_\odot$, or working in magnitudes.

Our own Galaxy is reckoned to contain some 10^{11} stars and, given that most stars have a lower output of light than the Sun, to have a luminosity around $2 \times 10^{10} L_{\odot}$. Taking the Sun to have a (blue) absolute magnitude of 5.48, this implies a total absolute magnitude for the Galaxy of $M_B \simeq -20.3$. Our giant neighbour M31 has $M_B \simeq -20.8$ (i.e. it is about 1.5 times as luminous), and many of the other well-known galaxies, such as M51, also have similar intrinsic brightnesses. Indeed, it was originally thought that the luminosity range of galaxies was quite small, with the Magellanic Clouds representing the faint tail of the distribution. (They have $M_B \simeq -18$ and -16.5.)

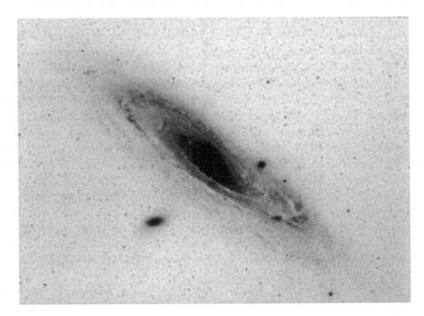

Figure 2.2   The giant spiral M31 with its dwarf elliptical companions M32 and NGC 205. (Courtesy: Mount Palomar Observatory.)

However, a little thought will suggest that this might be what is called a 'selection effect', rather than the true state of affairs. Very luminous galaxies may indeed be rather rare, but at the other end of the scale we might be biased against including low intrinsic luminosity systems in our samples, since they will need to be very nearby in order to *look* bright to us. On the other hand, very luminous galaxies will be seen easily throughout some large volume of space. Exactly this problem is encountered with stars. The bright stars in the night sky, like Vega and Rigel, contain a preponderance of (quite distant) stars intrinsically much brighter than the Sun. However, in a representative *volume* of space most stars would be relatively inconspicuous ones much less luminous than the Sun.

Some fairly small galaxies had been known for a long time, such as the companions to the Andromeda spiral, M32 and NGC 205. These are prototypes of the class of dwarf elliptical (dE) galaxies; their small size and low luminosity relative to M31 is evident even without any sophisticated measurements due to their close proximity to it (and equal distance from us). They are roughly 100

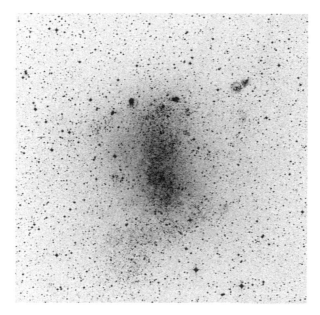

Figure 2.3   The Local Group irregular galaxy NGC 6882. (Courtesy: UKST image, © AAO.)

times less luminous than M31 itself. As their name suggests, dEs look rather like small versions of a normal (giant) elliptical galaxy, ellipsoidal in shape and with little or no internal structure. However, some dwarf ellipticals, denoted dE, N for 'nucleated', contain bright central star clusters like a tiny bulge component. Small irregular galaxies also exist, with similar luminosities, and unsurprisingly these are called dwarf irregular (or dI) galaxies. A nearby example is NGC 6882, one of the galaxies in which Hubble first found Cepheids (and the first one for which his results were actually published). None of these dwarf types is included properly in the tuning fork classification.

However, the first real evidence for a very wide range of galaxy luminosities came in the late 1930s, when Shapley announced the discovery of 'dwarf spheroidal' galaxies; in the first instance the dwarf spheroidal companions of our Galaxy known as the Fornax and Sculptor Dwarfs (named after the constellations in which they appear on the sky). Dwarf spheroidals (dSph) appear to be the extension to yet lower luminosities of the dEs, with Sculptor only 1/100 of the brightness of previously known systems. Since Shapley's day, many further dSph galaxies have been discovered around our Galaxy and around M31, their detection usually coming in the wake of improved astronomical technology and more thorough surveys of the sky such as the Palomar Observatory Sky Survey (POSS) in the 1950s and the UK Schmidt Telescope (UKST) sky surveys undertaken from the 1970s onwards. The lowest luminosity currently known companions to our Galaxy are the Draco and Ursa Minor dwarf spheroidals, while M31's most recently added companion And IX, discovered in 2004, appears to be marginally fainter and thus the least luminous galaxy known. Each is about 10^{-5} of the luminosity of M31 (i.e. a few $\times 10^5 L_\odot$).

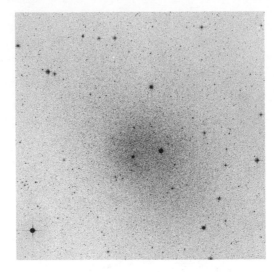

Figure 2.4   The Fornax dwarf spheroidal galaxy, discovered by Shapley in 1938. (Courtesy: UK Schmidt Telescope Unit.)

In absolute magnitude terms, these later discoveries thus range down to $M_B \simeq -8$, similar to the magnitudes of some very bright single stars. On the other hand, the brightest giant ellipticals in galaxy clusters, called cD galaxies, can range up to about $M_B \simeq -24$, that is, 25 times *brighter* than our Galaxy, in extreme cases.

Determining its intrinsic luminosity is relatively straightforward once we know the distance to a galaxy, but sizes are somewhat more problematic. Galaxies do not have obvious 'edges' any more than a gas cloud or the Earth's atmosphere has a definite edge – they simply fade away gradually. So, in order to describe the physical size of a galaxy, we need to consider how the stars (or other constituents) are distributed as a function of distance from the galaxy's centre. As far as stars are concerned, this is obviously reflected in the radial distribution of the light, that is, the surface brightness or amount of light per unit emitting area. In fact, the different types of galaxy (and their subcomponents) each have characteristic surface brightness profiles, $I(R)$ (section 2.4), but we can in general characterise the size of a galaxy by its 'half-light' or 'effective' radius. As the first of these terms implies, this is the radius of the circle (projected on the sky) which encloses half the total light of a galaxy. A large spiral galaxy like our own will have an effective radius R_e approaching 10 kpc. Large elliptical galaxies are of similar dimensions, while cD galaxies, which are surrounded by large extended envelopes, can have R_e up to $\sim$100 kpc. Dwarf galaxies are much smaller, with dwarf spheroidals ranging down to only 200 pc or so.

2.3   Surface brightness

Giant, high luminosity galaxies also have (at least in their central regions) high surface brightness, that is a high light output per unit area. This lends them their impressive appearance relative to the night sky. Dwarf galaxies, on the other hand, are typically of low surface brightness. Thus they look rather diffuse and 'woolly', and can be hard to see against even the faint sky foreground glow from our night-time atmosphere. Indeed, this was one of the main reasons for their relatively recent discovery. Physically this low surface brightness is due to the low surface density of stars. (Whether this also corresponds to a low *volume* density depends on how 'thick' the galaxy is along the line of sight.)

Surface brightness, as we might expect by now, is not usually quoted in the standard units of Wm^{-2}, but rather in $L_\odot \, pc^{-2}$ or magnitudes per square arcsecond. We can translate between these by noting first that $1L_\odot = 3.86 \times 10^{26} \, W$ and that $1 \, parsec = 3.086 \times 10^{16} \, m$. Thus

$$1L_\odot \, pc^{-2} = 4.05 \times 10^{-7} \, Wm^{-2} \qquad (2.3.1)$$

Also, the (blue) absolute magnitude of the Sun is $M_B = 5.48$, and as 1 pc obviously subtends 0.1 radians (20626 arcsecond) at the standard distance of 10 pc used in the calculation of absolute magnitudes, it follows that $1L_\odot \, pc^{-2}$ must also correspond to a surface brightness in magnitudes

$$\mu_B = M_{B\odot} + 5 \log (20626) = 27.05 \, B \text{ magnitudes per square arcsecond} \qquad (2.3.2)$$

(The latter units are sometimes written as $B\mu$).

If we characterise the overall surface brightness by the mean value inside the effective radius (i.e. $L/2\pi R_e^2$), then it turns out that (because of their comparable sizes) both types of giant galaxy have typical surface brightnesses of order $100L_\odot \, pc^{-2}$, or around $22 \, B\mu$. We can note, too, that not entirely coincidentally, as we are inside a giant spiral looking out, the brightness of the night sky is also about $22 \, B\mu$, or a little fainter at a really dark observatory site. Since ellipticals have much more steeply rising intensity profiles as we approach the centre than do spiral discs (Figure 2.5), the observed *central* values differ widely; just a few hundred $L_\odot \, pc^{-2}$ for discs but $\sim 10^4$ or more for ellipticals and bulges.

Dwarf galaxies are typically a factor 10 lower in average surface brightness than giants, and some are even fainter. Indeed, there is a whole class (or rather mixture of classes) known generically as 'low surface brightness galaxies' (LSBGs). These include the fainter dI galaxies (also known as Im IV and Im V galaxies, with the Roman numerals continuing the luminosity class idea of van den Bergh), as well as the dE and dSph galaxies and a more recently detected class of large but dim disc galaxies sometimes referred to as 'Malin 1 cousins' after their prototype. Malin 1 itself is a huge disc, of order 100 kpc in extent, but with a surface brightness, even in its centre, of only around $2L_\odot \, pc^{-2}$. We should note that not all dwarf galaxies have particularly low surface brightness, the trend being bucked, for instance, by the so-called 'blue compact dwarfs' (BCDs) which are precisely what their name implies and contain small, blue, high surface brightness central regions. As recently as 2000, another class of very small galaxies, ultra-compact dwarfs (UCDs) which appear to have properties between normal dwarfs and GCs, was added to the inventory.

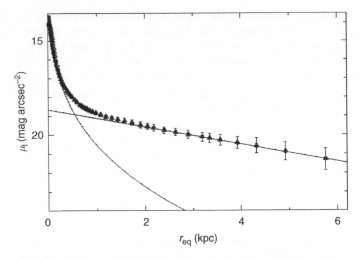

Figure 2.5   The I band radial surface brightness profile for NGC 3227 (Reproduced with permission from Sanchez-Portal *et al.*, 2004). Curves are for a pure exponential, pure $R^{1/4}$ law and a combined bulge plus disc profile.

2.4   Surface brightness profiles

Galaxies clearly have distributions of stars which peak in the middle. The way that the stars – and therefore the light from them – are distributed radially about the centre is then really what gives each different type of galaxy its characteristic appearance; we will consider the distribution of other constituents of galaxies in later chapters.

Spiral galaxy discs seen face-on have a particularly simple azimuthally averaged[1] surface brightness distribution which can be fitted by an exponential intensity profile.

$$I(R) = I_0 \exp(-R/a) \tag{2.4.1}$$

Here I_0 is the central intensity value and a is a scale length, describing how quickly the light drops off with radius R. In terms of surface brightness in magnitudes, this becomes

$$\mu(R) = \mu_0 + 1.086 R/a \tag{2.4.2}$$

The 1.086 factor – actually $2.5 \log_{10} e$ – arises from the logarithmic definition of magnitudes.

Ellipticals are traditionally fitted by the somewhat more complicated de Vaucouleurs or '$R^{1/4}$' law

$$I(R) = I_0 \exp\left(-(R/a)^{1/4}\right) \tag{2.4.3}$$

---

[1] The azimuthal averaging removes the effect of the spiral arms.

or

$$\mu(R) = \mu_0 + 1.086(R/a)^{1/4} \tag{2.4.4}$$

The bulge components of spiral galaxies are also often assumed to follow an $R^{1/4}$ law, but disentangling the two overlapping contributions is a difficult process observationally, and some workers in the field have suggested that bulges can be exponential or have some intermediate form.

Dwarf galaxies of both dE and dI varieties have profiles which are often close to exponential in form, but again there are suggestions that this may not be universal for dEs (which are spheroidal in their 3-D shapes, not flattened like spirals or irregulars). Specifically, the more luminous dEs may have profiles more like giant ellipticals than do the smaller dEs.

In any of these cases, we can see that the scale length a represents the radial distance at which the intensity has dropped by a factor e $(=2.718)$, or just over 1 magnitude in surface brightness.

We can also determine the total luminosities of galaxies in terms of these profile parameters. Indeed we merely need to calculate the integral

$$L_{\mathrm{T}} = \int_0^{\infty} 2\pi R I(R)\mathrm{d}R \tag{2.4.5}$$

where for simplicity we assume that the image of the galaxy is circular on the sky ('face-on' to the observer). More generally, we can calculate the 'curve of growth', the luminosity contained within a circle of radius R projected on the sky,

$$L(R) = \int_0^R 2\pi w I(w)\mathrm{d}w \tag{2.4.6}$$

For exponential profiles we quickly obtain

$$L(R) = 2\pi a^2 I_0(1 - (1 + x)e^{-x}) \tag{2.4.7}$$

where $x = R/a$. Thus, letting x tend to infinity

$$L_{\mathrm{T}} = 2\pi a^2 I_0 \tag{2.4.8}$$

$L(R)$ also provides the means to determine another useful measure of the radial scale of a galaxy, the effective radius R_{e}, introduced above. Since this contains half of the total light of the galaxy (as projected on the sky), it obviously requires

$$L(R_{\mathrm{e}}) = 0.5L_{\mathrm{T}} \tag{2.4.9}$$

so $(1 + x)e^{-x} = 0.5$, giving $x = 1.69$. Thus

$$R_{\mathrm{e}} = 1.69\,a \tag{2.4.10}$$

The surface brightness at R_e, written as I_e or μ_e, can be used as an alternative to I_0 or μ_0 in describing the typical surface brightness of a disc. It is easy to see from above that the relationships between the various parameters will be

$$I_e = I_0\, e^{-1.69} = 0.184\, I_0 \qquad (2.4.11)$$

$$\mu_e = \mu_0 + 1.83 \qquad (2.4.12)$$

and

$$L_T = 3.81 \pi R_e^2 I_e \qquad (2.4.13)$$

For ellipticals (or bulges) the integrals are lengthier to compute, giving

$$L_T = 8!\pi a^2 I_0 \qquad (2.4.14)$$

$$L(R) = L_T \left(1 - e^{-x} \sum_{k=0}^{7} x^k\right) \qquad (2.4.15)$$

where now $R/a = x^4$. Also

$$R_e = 7.67^4 a = 3461\, a \qquad (2.4.16)$$

$$I_e = e^{-7.67} I_0 \qquad (2.4.17)$$

$$\mu_e = \mu_0 + 8.33 \qquad (2.4.18)$$

and it is sometimes convenient to write the profile as

$$I(R) = I_0\, \exp\left(-7.67(R/R_e)^{1/4}\right) = I_e\, \exp\left(-7.67\left[(R/R_e)^{1/4} - 1\right]\right) \qquad (2.4.19)$$

Finally

$$L_T = 7.22 \pi R_e^2 I_e \qquad (2.4.20)$$

2.5   Apparent sizes

The apparent size of a galaxy on the sky must depend on its profile parameters, too. Which part of the galaxy we can actually see will depend on its contrast against the night sky and the 'noise' in the observations. Even for otherwise perfect imaging, we are still limited by the fact that we detect a finite number of photons; so our measurement of a galaxy or part of a galaxy which emits, say, n_g photons can never be more accurate than the 'Poisson noise' of $\sqrt{n_g}$. In addition, there will be noise from the number of photons received from the underlying sky, n_s, so in fact the 'signal-to-noise ratio' will be (at best)

$$\frac{S}{N} = \frac{n_g}{(n_g + n_s)^{1/2}} \tag{2.5.1}$$

At some point, the number of galaxy photons in a given area (e.g. a detector pixel) will drop to the point where S/N is so small (say 3) that you cannot distinguish it reliably from the background. This is the isophotal detection threshold, which we can represent either by a minimum intensity level I_{min}, typically a few per cent of the sky background level, or a corresponding limiting surface brightness, μ_{lim}. The size of the galaxy out to this limit is then its 'isophotal diameter'. Similarly, of course, we are not sensitive to the remaining total light outside this isophote, so a more directly measurable quantity than L_T is the isophotal luminosity L_{iso} which only integrates the light out to the isophotal radius.

To obtain L_T from observation, one must therefore extrapolate the measured brightness. If we believe that an exponential or $R^{1/4}$ law remains a good description of the light profile even in the outer parts, then we can simply use the observed values of a and I_0 and obtain the theoretical value of L_T. Alternatively, one can determine a radius which should contain a very large fraction of the total light, and sum the observed data out to that point (which may be well beyond the isophotal size). Two schemes are in common use. First, there is the Kron radius, which basically comes from a second moment of the light distribution

$$R_K = \frac{\int R^2 I(R) \, dR}{\int R I(R) \, dR} \tag{2.5.2}$$

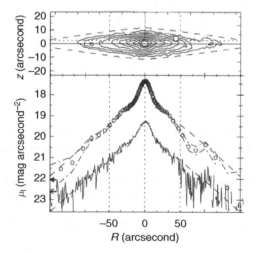

Figure 2.6 Observed profiles showing the effect of noise in limiting the maximum detectable size of an image (Reproduced with permission from Kregel, van der Kruit and de Grijs, 2002).

Going out to $2.5R_K$ will encompass over 90% of the light for any realistic galaxy profile. Second, there is the Petrosian radius which is defined as the radius at which the surface brightness drops below a certain fraction (usually 0.2) of the average surface brightness out to that point, i.e.

$$I(R_P) = 0.2\frac{L(<R_P)}{\pi R_P^2} \qquad (2.5.3)$$

Choosing a suitable multiple of R_P, we can again construct a radius which will contain a very large fraction of the galaxy's light for any sensible profile shape. In fact, with the above definition, $2R_P$ encloses 98% of the light for an exponential profile, 95% for an unresolved (star-like) profile and over 80% for an elliptical galaxy profile. Working the other way, radii such as this can also be used for crude estimates of morphological type, based on the concentration of an image. For instance, if R_N is the radius containing N% of the light inside $2R_P$, then $C_P = R_{90}/R_{50}$ takes a value around 2.3 for disc galaxies and 3.3 for ellipticals.

2.6   The luminosity function

The distribution of luminosities amongst a representative collection of galaxies is an important datum both observationally, as it gives essentially a census of galaxies of the different varieties, and theoretically, where, due to its relationship to the distribution of galaxy masses, it provides a measure of the amount of structure formed in the Universe by the present day. Generally speaking, the numbers of very bright galaxies become decreasingly small, but as we go to lower luminosities we find large numbers of objects. To provide a precise quantification of this, we define the 'luminosity function', ϕ, to be the number of galaxies per unit volume (in practice per Mpc^3) per unit luminosity interval.

The most common parameterisation of the luminosity function (LF) is that proposed by Paul Schechter in 1976, which we can write as

$$\frac{dn}{dL} = \phi(L) = \frac{\phi_*}{L_*}e^{-L/L_*}\left(\frac{L}{L_*}\right)^{\alpha} \qquad (2.6.1)$$

Here ϕ_* represents a normalisation factor (loosely the number density of bright galaxies) and L_* is a characteristic luminosity where the numbers level off from the steep exponential cut-off at the bright end to the power-law of slope α at the faint end.

In terms of magnitudes, this can alternatively (but more awkwardly) be written as the number density of galaxies per unit volume in a one magnitude interval, giving

$$\phi(M) = 0.4\log_{10}e\;\phi_*\exp\left(-10^{-0.4(M-M_*)}\right)10^{-0.4(\alpha+1)(M-M_*)} \qquad (2.6.2)$$

where M_* is the magnitude corresponding to luminosity L_*. To obtain this form we have used the usual identity

$$M - M_* = -2.5 \log (L/L_*) \tag{2.6.3}$$

and the fact that we must have

$$\phi(L)dL = \phi(M)dM \tag{2.6.4}$$

if dM is the magnitude range corresponding to luminosity interval dL.

Current estimates in the B band give

$$\phi_* \simeq 0.0055 \, \mathrm{Mpc}^{-3},$$

$$L_* \simeq 2 \times 10^{10} L_\odot$$

or

$$M_* \simeq -20.6$$

(close to the brightness of our Galaxy), and

$$\alpha \simeq -1.2$$

In terms of magnitudes, the latter value implies that the numbers of galaxies per one-magnitude interval rise slowly towards fainter objects: if α were exactly -1, there would be the same number of galaxies in each magnitude bin at the faint end. However, it may be that we should consider the dwarfs separately, and that the slope for the dwarfs steepens to something like $\alpha = -1.5$. Note that the slope cannot be as steep as $\alpha = -2$, at least not for long, otherwise the total amount of light from the whole galaxy population would diverge.

We might also note here that in the Schechter parameters we have modified the values given in the original literature sources for differences in the assumed Hubble constant. For instance, M_* is measuring a brightness via the inverse square law or in fact, since it is in magnitudes, via the distance modulus equation (1.6.3). Thus each calculated M for a galaxy in a survey will vary as $-5 \log D$, and therefore for a given observed redshift as $+5 \log H_0$ (from equation 1.5.1). M_* will then vary the same way. It became standard to parameterise our ignorance of the true value of H_0 via $h = H_0/100 \, \mathrm{km \, s^{-1} \, Mpc^{-1}}$. Thus the calculated M_* has an 'uncertainty' term of $5 \log h$, being brighter (more negative) for smaller values of h. Similarly ϕ_* is a volume density, so will be inversely proportional to assumed distances cubed and therefore to h^3. It is thus smaller for smaller h. As H_0 now seems to be tied down quite solidly, we have dispensed with this uncertainty and quote all our values for an assumed $H_0 = 70 \, \mathrm{km \, s^{-1} \, Mpc^{-1}}$, that is $h = 0.7$.

Writing $\mathcal{L}$ for the total luminosity density,

$$\mathcal{L} = \int_0^\infty L\phi(L)dL \tag{2.6.5}$$

For a pure (single component) Schechter function, this becomes

$$\mathcal{L} = \phi_* L_* \Gamma(2 + \alpha) \tag{2.6.6}$$

Here $\Gamma(x)$ is the gamma function, which cannot be written in simpler terms (unless x is an integer, in which case it is the same as the factorial function $(x-1)!$), but can be looked up in mathematical tables. Notice that if α were exactly -1, we would have $\Gamma(2+\alpha) = \Gamma(1) = 1$, so $\mathcal{L} = \phi_* L_*$. With the more accurate numbers given above, we obtain

$$\mathcal{L} \simeq 1.4 \times 10^8 L_\odot \, \mathrm{Mpc}^{-3} \tag{2.6.7}$$

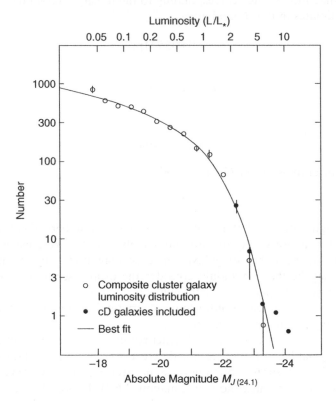

Figure 2.7 A typical Schechter function fit to the galaxy luminosity function (Reproduced with permission from Schechter, 1976).

The total *number* of galaxies per unit volume is similarly given by

$$n = \phi_* \Gamma(1+\alpha) \tag{2.6.8}$$

which is generally a divergent function. However, in reality, galaxies do not exist with arbitrarily small luminosities. If, instead, we want the number of galaxies brighter than some minimum brightness L_{min}, then we obtain

$$n(>L_{\mathrm{min}}) = \phi_* \Gamma(1+\alpha, L_{\mathrm{min}}/L_*) \tag{2.6.9}$$

where the relevant function now is the 'incomplete gamma function'.

2.7 Redshift surveys

In order to ascertain the form of the luminosity function, we obviously need to know the absolute brightnesses of a large number of galaxies. Observing their apparent brightnesses is relatively simple, so the key is obtaining distances to large numbers of galaxies. Even today, it is still very hard to determine distances to galaxies directly. Fortunately, as we have seen in Chapter 1, the cosmological expansion of the universe provides us with a practical alternative. By Hubble's Law we know that the redshift of a galaxy will be proportional to its distance, so we can use redshift as a substitute for cosmic distance. Thus determining the LF requires us to measure redshifts for large – and complete – samples of galaxies. By complete, here, we mean that all the galaxies down to some limiting brightness in some representative region of sky have been observed. Clearly if we missed out some objects then our LF could be in error. Such an undertaking is known as a 'redshift survey' and over the last two decades technological advances have meant that sample sizes have risen from an order of a hundred galaxies to the hundreds of thousands of objects observed by current 'state of the art' redshift surveys. The 2dF galaxy redshift survey (2dFGRS), for instance, has obtained redshifts for around 220 000 galaxies. It uses the '2 degree Field' multi-object spectrograph on the 3.8-m Anglo-Australian Telescope in New South Wales, which can obtain 400 spectra simultaneously. Similarly, the Sloan Digital Sky Survey (SDSS) is a US project to obtain photometry (i.e. measured brightnesses) and spectra for around 1 million objects. This has been carried out using a purpose-built 2.5-m telescope at Apache Point in New Mexico.

To calculate an LF from such a survey, we need first to determine the absolute magnitude of each galaxy in the sample. This is rather more complicated than one might initially expect. For instance, unless the galaxies are sufficiently 'local', we need to consider the cosmological curvature of the universe when calculating distances from redshifts via Hubble's law. We will deal with this aspect of the problem more fully in Chapter 8. But even after we have obtained a correct distance, there are still several effects to take into account.

2.7.1 Galactic extinction

For a start, we need to allow for the amount of foreground absorption by the dust in our Galaxy. This will depend on where the galaxy is on the sky, with the absorption being greater the nearer it is to the Galactic Plane, since we are then looking through a greater depth of interstellar medium in this direction. As can be seen from Figure 2.8, if we approximate our Galaxy by a uniform slab with the Sun in the central plane, then the line-of-sight thickness will vary as cosec b, where b is called the Galactic latitude in analogy with geographical latitude.

Now, each element of length will remove a certain *fraction* of the light; so if we write $I(x)$ for the intensity of the light after it has passed through a length x of the interstellar medium (ISM) from its source, then after a further distance Δx we must have

$$I(x + \Delta x) = I(x) - \Delta I = I(x)[1 - \kappa \Delta x] \tag{2.7.1}$$

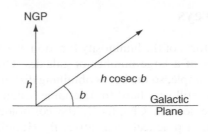

Figure 2.8   The cosecant law for path length through the Galactic disc (taken to be a slab with half-thickness h). Here NGP indicates the direction to the North Galactic pole.

where κ represents the rate at which the light is absorbed (called the absorption coefficient). Thus we can write the differential equation

$$\frac{\mathrm{d}I}{\mathrm{d}x} = -\kappa I \tag{2.7.2}$$

which has the solution

$$I = I_c \, e^{-\kappa x} \tag{2.7.3}$$

if I_c was the original unobscured intensity at $x = 0$. This is usually written

$$I = I_c \, e^{-\tau} \tag{2.7.4}$$

where τ is called the optical depth. Translating into our usual magnitude system, this becomes

$$m = m_c - 2.5 \log (e^{-\tau}) = m_c + 1.086\tau \tag{2.7.5}$$

or

$$m = m_c + A \tag{2.7.6}$$

where m_c is the magnitude we would see in the absence of absorption by the dust, and A is now called the (total) 'extinction'.

A rough estimate of the absorption in, say, the blue band when looking all the way out of the Galaxy is $A_B \simeq 0.2 \operatorname{cosec} b$ magnitudes. In practice, of course, the Galaxy is not uniform and the Sun is not at the centre, so we will also expect some variation with the angle around the plane – the Galactic longitude ℓ, measured from $0°$ towards the Galactic Centre (Figure 2.9). As the absorption is due to dust, we can deduce A_B as a function of ℓ and b by looking at a map of far-infra red (FIR) dust *emission*, as is shown in Figure 2.10. For later use, we can also note that the 'selective' extinction – also called the 'reddening' – is the difference in extinction at two different wavelengths, for instance in the B and V bands, written $E(B-V)$, which will change the apparent colour of obscured objects.

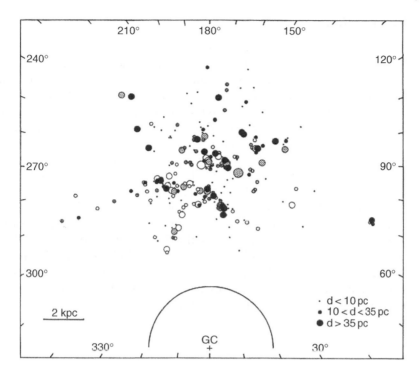

Figure 2.9 A map of the region of the Galaxy around the Sun (Reproduced with permission from Crampton and Georgelin, 1974) illustrating the orientation of Galactic longitudes (The Sun is at the centre of the plot and the longitudes are marked around the edge).

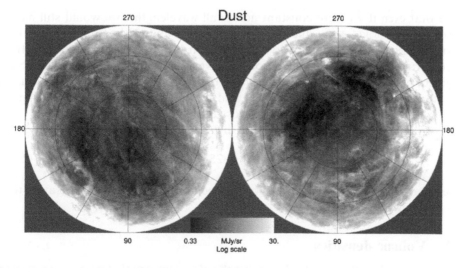

Figure 2.10 Far-infra-red dust emission map (from Schlegel, Davis and Finkbeiner, 1998). This same dust is responsible for the interstellar extinction.

2.7.2   k-corrections

Next we need to consider another effect of the cosmological redshift. Clearly the whole of the spectrum is shifted to longer wavelengths, not just individual spectral lines. Thus, when we observe with a certain detector, or through a certain filter, which admits light of only a certain range of wavelengths, it will be sensitive to different *emitted* wavelengths depending on the redshift of the galaxy being observed. For instance, if we observe with a blue filter, this will typically have a passband between 370 and 490 nm. For a very nearby galaxy, obviously the light emitted in this same range 370–490 nm will be detected. However for a galaxy at redshift $z = 0.1$, wavelengths are shifted by a factor 1.1; so the *restframe* wavelengths emitted by the galaxy, which will eventually be detected, are those between 370/1.1 and 490/1.1, that is about 336 and 445 nm. If, as will almost certainly be the case, the galaxy emits a different amount of energy in these two ranges, then in order to compare like with like, we need to correct the flux observed from the redshifted object for this difference. This is called the 'k-correction' (for historical reasons).

More formally, if we let $F_\lambda(\lambda)$ be the flux from a certain galaxy at wavelength $\lambda - F_\lambda$ is called the galaxy's spectral energy distribution (SED) – then the ratio of the fluxes detected in the redshifted and un-redshifted bandpasses will be

$$K = \frac{F_\lambda(\lambda_B/(1+z))\Delta\lambda_B/(1+z)}{F_\lambda(\lambda_B)\Delta\lambda_B} \qquad (2.7.7)$$

where λ_B is the central wavelength of the B passband (in this case) and $\Delta\lambda_B$ is the bandwidth. The k-correction is traditionally expressed in magnitudes, so is therefore

$$k(z) = -2.5 \log K = 2.5 \log(1+z) - 2.5 \log\frac{F_\lambda(\lambda_B/(1+z))}{F_\lambda(\lambda_B)} \qquad (2.7.8)$$

Notice that even if F_λ were constant across all wavelengths, we would still have a k-correction $2.5\log(1+z)$, due to the so-called bandwidth stretching; that is, a narrower range of restframe wavelengths contributes to the observed flux from the redshifted galaxy. In this calculation we have neglected the variation of F_λ across the given bandpass and the efficiency of the detector in collecting photons of different wavelengths, but we can easily fold these into the calculation by using an appropriately weighted integral of F_λ across the band.

Thus, finally we have that the effective distance modulus will be

$$m - M = 5 \log D(z) + 25 + k(z) + A(\ell, b) \qquad (2.7.9)$$

where the distance to the galaxy is now assumed to be measured in Mpc (rather than pc as in the usual definition).

2.7.3   Volume densities

In order to calculate the galaxy LF, we can therefore suppose that we have a complete sample of galaxies, all with measured z and m from which we can deduce M for each

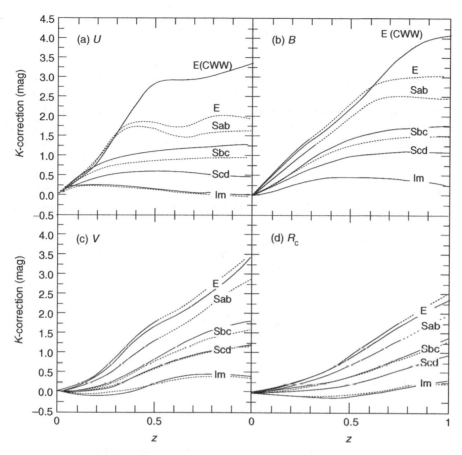

Figure 2.11 k-corrections in the *U, B, V* and R_c bands for various galaxy types (Reproduced with permission from Fukugita, Shimasaku and Ichikawa, 1995).

galaxy. But the distribution of the *M* values is still not the LF, for the reason alluded to in section 2.2, viz. more luminous galaxies are visible at much greater distances than less luminous ones. To allow for this, we *weight* each galaxy by the inverse of the volume in which it could lie and still appear bright enough to get into our sample. If our sample has an apparent magnitude limit m_{lim}, we can easily calculate the maximum redshift at which a galaxy of given absolute magnitude *M* could be seen from

$$m_{\text{lim}} - M = 5 \log D(z_{\text{max}}) + 25 + k(z_{\text{max}}) + A(\ell, b) \qquad (2.7.10)$$

The available volume will then be

$$V_{\text{max}} = V(z_{\text{max}}) \qquad (2.7.11)$$

where again we may need to account for the curvature of space when calculating the volume. Ignoring this complication, we would have simply

$$V_{\max} = (\Omega/3)D_{\max}^3 = (\Omega/3)(cz_{\max}/H_0)^3 \qquad (2.7.12)$$

where Ω is the size of the survey area on the sky in steradians.

Thus, finally we can obtain the volume density of galaxies of a given absolute magnitude M by summing up all the contributions $(1/V_{\max})$ for galaxies of that M. (Despite the apparent complexity of this last step, it is easy to see that it makes sense by considering the case of just a single galaxy). The weighting allows for the different volumes in which galaxies of different luminosities can be seen, but of course we still actually see far fewer dwarfs in any such sample than we do giants. Thus the (small) number of dwarfs we count is relatively insensitive to the actual form of the LF (Figure 2.12) so, going the other way, there will be much greater statistical uncertainty at the faint end of the LF than at the bright end. This has led to considerable

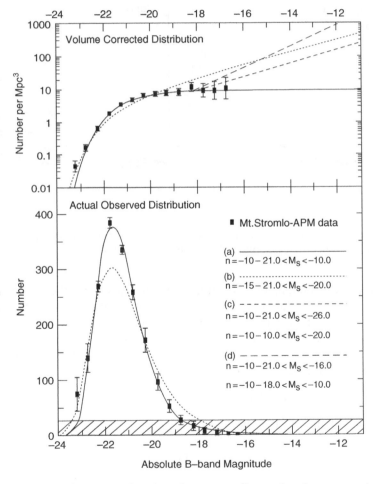

Figure 2.12 Different luminosity functions (upper panel) can give rise to very similar numbers of observed dwarf galaxies (lower panel) (Reproduced with permission from Driver and Phillipps, 1996).

argument over the years as to the value of the faint end slope α and, indeed, whether a single Schechter function remains a good fit to the LF over its whole range.

This uncertainty has been increased by the fact that many low luminosity galaxies are also of low surface brightness and not all surveys may be equally sensitive to such objects. Even ordinary imaging surveys will have a limit on the lowest surface brightness galaxies that they can detect above the overall brightness of the night sky (which is 10–100 times higher than that of many dwarf galaxies, remember) and redshifts are much harder to obtain, still.

We should also note that when comparing our observations with theoretical ideas on structure formation, we may often be more interested in the *mass* function of galaxies rather than the luminosity function. This then requires us to allow for the (rather self-explanatory) 'mass-to-light ratio' that we met in Chapter 1. The masses of galaxies will be treated in detail under their specific types.

2.8   Galaxies at all wavelengths

So far, we have been thinking mainly of the optical light output from galaxies. In particular, we have been treating galaxies as if they contain only stars (and obscuring dust). However, they emit radiation at other wavelengths, too. Indeed some, like the radio galaxies we met briefly in Chapter 1, are more prominent when viewed in other parts of the electromagnetic spectrum. We should not assume that the optical appearance is the whole story by any means. Modern astrophysics utilises all wavelengths of radiation, from gamma-rays and X-rays, through the ultra-violet, optical and infra-red, up to the radio, and the study of galaxies is no exception in this regard.

To conclude our introduction to the various types in the galaxy zoo, we should therefore add to the inventory those objects particularly associated with non-optical emission. Starting at long wavelengths, most galaxies in fact show radio emission at some level. Spiral galaxy discs, for instance, have radio emission associated with star formation and with interstellar matter. However, radio galaxies, as such, are giant elliptical galaxies. These are powered by 'central engines' containing massive black holes and often exhibit jets and twin lobes of emission on either side, in some cases extending for hundreds of kpc, but containing much smaller ($\sim$kpc sized) 'hot spots'. The total power outputs can range from 10^{34} to 10^{38} W.

At the shorter radio wavelengths, the millimetre and sub-millimetre ranges have only recently been opened up technologically. At these and the adjoining FIR wavelengths, we see primarily thermal re-emission from dust. From Wien's law for blackbody radiation,

$$\lambda_{\text{peak}} T = 2.898 \times 10^{-3} \, \text{m K} \qquad (2.8.1)$$

we can see that the relevant temperatures for objects whose spectra peak at these wavelengths are less than 100 K. Indeed, the most recent observations at the longer wavelengths have emphasised the important role of dust at only 10–20 K. The FIR region is virtually impossible to access from the ground as photons of these wavelengths are strongly absorbed by the Earth's atmosphere. Thus this regime is the

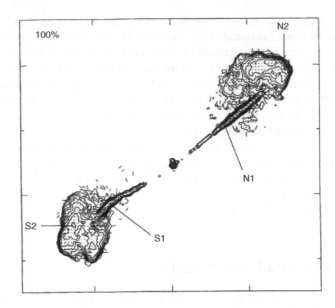

Figure 2.13 A radio map of the radio galaxy 3C438 (Reproduced with permission from Hardcastle *et al.*, 1997), showing the jets and lobes.

province of space observatories and was first properly probed by the Infra-Red Astronomical Satellite (RAS) in the 1980s. These observations led to the discovery of a new variety of galaxy, or at least of galaxies in a new guise. These are the ultra-luminous infra-red galaxies (ULIRG), powered by starbursts, that is the rapid formation of large numbers of stars over a short time period. The starburst regions are enshrouded in thick dust layers, hence their prominence at infra-red rather than optical wavelengths; they can put out more than $10^{13} L_\odot$ just in the FIR. High redshift versions of these are prominent in the sub-mm wavelength range, which can now be accessed from the ground using instruments like the Sub-millimetre Common User Bolometer Array (SCUBA) on the James Clerk Maxwell Telescope on Mauna Kea in Hawaii.

At short wavelengths, on the other hand, we are probing completely different parts of galaxies and seeing the results of highly energetic processes. X-rays, in particular, are produced thermally when material reaches temperatures of 10^6 K or more. After pioneering rocket flights, X-ray astronomy as such really began with the Uhuru satellite in the early 1970s and continued with increasing sophistication with the space observatories Einstein and ROSAT in the 1980s and 1990s. Three basic types of source are seen. Bright stellar sources in our own and other galaxies are primarily X-ray binary systems, where the X-rays are produced when material lost from one star falls onto a compact companion. Hot gas in galaxies also produces X-rays, as does that between the galaxies in clusters. Finally, we have the nuclei of galaxies which harbour black holes – the so-called Active Galactic Nuclei or AGN – where again the emission is due to accretion of material onto a central object.

Diffuse Galactic emission in γ-rays, with energies of tens of MeV, was first seen by the SAS2 and COS B satellites, again in the 1970s. The Compton Gamma Ray

Observatory, launched in 1991, had the necessary resolution to detect individual distant sources, often AGN, and also discovered the strange gamma-ray bursters (GRBs), now thought to be even more energetic versions of supernovae in distant galaxies.

2.9   Active galaxies

Active Galactic Nuclei in fact radiate strongly at most wavelengths, though they may be largely absorbed in the optical, in some cases. The first AGN were discovered in the optical–ultra-violet region as intense blue emission regions in the centres of (almost always spiral) galaxies. These objects, called Seyfert galaxies after Carl Seyfert who first catalogued them in the 1940s, showed very broad emission lines of highly ionised atoms, suggestive of very high temperatures. In the 1950s active nuclei were also detected as radio sources, often in elliptical galaxies and often associated with jet-like features, as in M87.

Then in 1963 Schmidt identified the first quasars – quasi-stellar radio sources – as a new type of extragalactic object with strong emission lines in their spectra. In fact they are essentially even more extreme versions of the other AGN, now so powerful that in the optical they can outshine the whole of their host galaxy by a factor 10 or more. Later, similar objects, but without the radio emission, were discovered and became known as QSOs, for quasi-stellar objects (QSOs). Nowadays, though, both 'radio loud' and 'radio quiet' types are often interchangeably referred to as 'quasars' or 'QSOs' despite the origins of the terms. Another type of AGN we should also mention here are the BL Lac objects, named after their prototype BL Lacertae, a supposed variable star which turned out to be an AGN. These show no easily visible emission lines, but have the other characteristics of QSOs and are also very variable in their light output over very short time periods. Along with the class of 'optically violently variable' (OVV) quasars BL Lac objects are also referred to as Blazars. Optical variability is a very important characteristic of many quasars and provided the original evidence that their prodigious light output actually comes from a tiny region, perhaps no larger than the Solar System. This is because, roughly speaking, large changes in luminosity cannot be seen on periods of, say, a day, if the source is more than a light day across, as the light travel time across the source would 'wash out' the variations.

2.10   Galaxy environments

We have already seen that our Galaxy has a number of small companions or satellites such as the Magellanic Clouds, as well as a giant neighbour in M31. All of these galaxies, and around 30 others, make up what is called the Local Group. This group contains just three giant galaxies, the Galaxy, M31 and M33, the rest being irregulars (some reasonably large like the Magellanic Clouds, most rather smaller) or dwarf ellipticals and spheroidals. The latter occur as satellites of the larger galaxies (so are rather limited in their distribution), but the irregulars can also be 'free flying' and fill out the overall space of the Local Group. This is

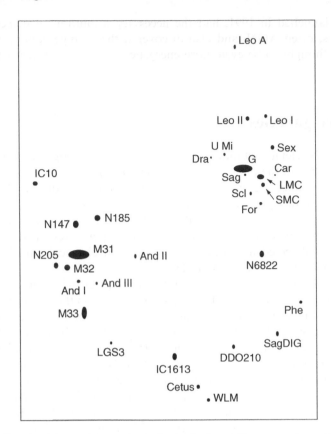

Figure 2.14   Sketch of the positions of galaxies in the Local Group, as viewed from 'above' its main plane. Our Galaxy is labelled 'G' and the plot is of an area roughly 1.2 by 1.5 Mpc.

conventionally taken to be a region of radius 1.5 Mpc around the mid-point between M31 and the Galaxy (Figure 2.14). Note that the projected map shown in the figure is actually quite a good representation of the real distribution, as the Local Group is rather flat, that is all the galaxies lie quite close to a central plane.

Beyond the Local Group, we find that most other nearby galaxies also lie in similar structures. Figure 2.15 shows our immediate surroundings, with the Sculptor or South Polar Group a few Mpc away in one direction and the M81 group a similar distance away to the other side, for instance. A little further away again, we come to the Cen A group, named for its largest member Centaurus A, the nearest giant elliptical galaxy to us and one of the brightest radio sources in the sky. Note that many of the groups seem to line up, making larger filamentary structures. Few galaxies are completely isolated.

If we move out to larger scales still, then we soon meet the Virgo Cluster in the north and the Fornax Cluster in the south. These are at distances around 18–20 Mpc. As with our neighbouring individual galaxies, the nearest clusters of galaxies are often named after the constellation in which they appear to lie on the sky. The fact that there is a concentration of nebulae in Virgo was already known to Messier and

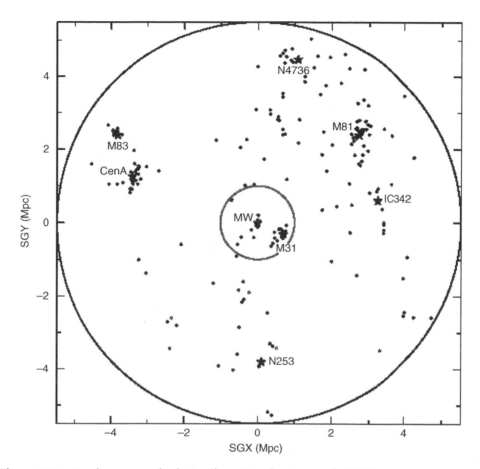

Figure 2.15   Nearby groups of galaxies (from Karachentsev *et al.*, 2003).

his contemporaries. Clusters contain hundreds to thousands of galaxies, but are much rarer than small groups. The Virgo Cluster is also the centre of an even larger structure, the Virgo Supercluster, of which the Local Group, the M81 Group and so on are outlying members. Indeed, the Local Group is essentially at the edge of the Virgo Supercluster region, and in the Fornax direction, for instance, there are virtually no galaxies between the Local Group and the Fornax Cluster itself. This is an example of a 'void' in the galaxy distribution. We will have more to say about clusters and voids in Chapter 7.

For now, we will mention only how the environment can influence general galaxy properties. For instance, it turns out that the denser the region, the greater is the fraction of elliptical galaxies. This so-called 'morphology–density relation', demonstrated by Alan Dressler in 1980, shows that in small groups and other low density regions (collectively called the 'field'), the fraction of bright galaxies which are spirals is around 80%, with about 20% S0s and very few ellipticals. As we move in from the outskirts of clusters, the fraction of spirals decreases steadily (to near zero by the

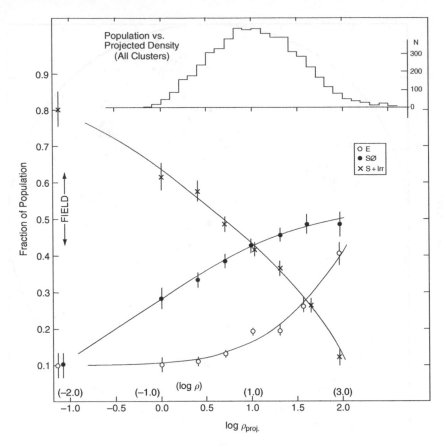

Figure 2.16   The morphology–density relation (Reproduced with permission from Dressler, 1980).

time we reach the densest regions), while the S0 fraction rises steadily (up to ~50%) but is rapidly caught up by the elliptical fraction once we reach high densities. The numbers of dwarf galaxies also appear to show a dependence on density, with their greatest prevalence (relative to giants) being at the densities seen in the outer parts of clusters, rather than the dense central regions or the low density field. The dE and dSph types occur primarily in clusters or, as in the Local Group, as satellites of giant galaxies, but appear to avoid general field regions. Notice that these differences also imply that the galaxy LF must be different in different environments. In particular, the greater number of dwarfs compared to giants seen in clusters is equivalent to a steeper faint end of the LF in clusters than the field.

Each of these relationships between density and morphology must presumably be due to either 'nature or nurture', that is, a result of different formation mechanisms or different evolutionary paths for galaxies in different environments. We will consider galaxy formation and evolution in detail in Chapter 8. We will also have more to say about the overall distribution of galaxies in the universe and what that may have to tell us about structure formation on the largest scales.

3 Elliptical and lenticular galaxies

3.1 Numbers

Early type giant galaxies, the ellipticals and lenticulars (or S0s), are distinguished by their fairly featureless morphologies, though of course S0s have both bulges and discs. As the classic 'galaxies as collections of stars' it was long thought that this simplicity in appearance should translate into relatively simple physical character-istics such as their dynamics and their formation and evolutionary processes. How-ever, in the last two decades or so this expectation has dissipated and it is now clear that their simple appearance is deceptive.

In terms of their environment, we know already from the morphology–density relation (section 2.10) that early type giants predominate in clusters. Indeed this is true in both senses; the majority of bright galaxies in clusters are early types *and* the majority of giant early type galaxies are in clusters. This is not mere tautology as only a small fraction (<10%) of all galaxies are in large clusters. Recall that in the inner parts of rich clusters as many as 80% of the large galaxies are early types, and right in the centres of densely packed clusters essentially all of them are. In the field, early types are distinctly uncommon, with S0s perhaps accounting for around 10–20% of the total number of giants. Those ellipticals like Cen A which are in small groups are in a distinct minority.

As for their total numbers, we can determine the luminosity function separately for early type galaxies, either from a large survey which properly samples both field and cluster regions, or by looking in detail at individual clusters of galaxies. In the former case, it is the selection of the early types which can be difficult, since

The Structure and Evolution of Galaxies Steven Phillipps
© 2005 John Wiley & Sons, Ltd

morphology can be determined reliably with ground-based data only for relatively nearby galaxies, as otherwise the images are too small for good discrimination. The limiting factor from the ground is the blurring due to the Earth's atmosphere (called 'seeing'), so an obvious answer is to use observations with the HST. However, the HST has a very small field of view (the size of the patch of sky that can be imaged in one exposure), so that it is not feasible to make large surveys in this way.

Until recently, this – and their general scarcity outside clusters – led to relatively uncertain determinations of the LF for early types, though observations suggested relatively few lower luminosity early types, implying a declining LF at intermediate to faint magnitudes (i.e. a faint end slope $\alpha > -1$). Using the newer large-redshift surveys with the galaxies differentiated by morphological type, we find that the Schechter function parameters for (non-dwarf) early types are approximately:

$$M_{B*} \simeq -20.4$$

$$\phi_* \simeq 0.0034$$

$$\alpha \simeq -0.5$$

We can notice three things about this. First, the characteristic magnitude M_* is marginally fainter for Es and S0s than for all galaxies combined ($M_{B*} \simeq -20.6$). Second, the volume density is obviously lower, but ϕ_* is only a factor 1.6 less than that for the overall galaxy population. However, note that this is not the same as saying that around 60% of all galaxies are early type giants. This will be approximately true around M_* (and at very bright absolute magnitudes nearly all the galaxies are early types), but the third point that the faint end slope parameter is larger (i.e. less negative) means that their numbers per magnitude interval do not increase towards faint magnitudes like those of galaxies in general. Thus the overall fraction of galaxies which are giant Es and S0s is rather small.

Recently, the 2dFGRS (section 2.7) has also been used to define type-specific LFs from its huge galaxy sample. However, the input data to the catalogues used was photographic and of limited resolution for trying to distinguish the morphologies of galaxies down to the magnitude limit at $m_B \simeq 19.5$. Thus the selection of early type galaxies was based on the spectra used for the redshift determinations. They will therefore be similar, but probably not identical, to samples selected by image morphology. Fortunately, the agreement seems quite good.

As noted above, an alternative route to the LF is to concentrate on single galaxy clusters. Here one can simply count the apparently early type galaxies in a certain area of sky containing the cluster (as a function of their magnitudes) and subtract off the equivalent numbers seen in another patch of sky outside the cluster. This should account for the typical number of galaxies seen in front of or behind the cluster. This type of statistical approach, requiring no individual redshifts (though we still need to know the redshift of the cluster), can provide reasonably accurate LFs down to faint absolute magnitudes. Observations of the Virgo cluster and a number of more distant rich clusters seem to confirm the shape of the early type LF given above, though of course we cannot obtain ϕ_*, the overall average space density, this way.

3.2 Surface brightness laws

The declining LF at intermediate magnitudes suggests that giant and dwarf elliptical galaxies (the latter appearing in increasing numbers at faint magnitudes) may be separate populations, and this is also indicated by their surface brightnesses. As we saw in section 2.4, giant ellipticals have a characteristic de Vaucouleurs or $R^{1/4}$ law radial profile.

$$I(R) = I_0 \exp\left(-(R/a)^{1/4}\right)$$

or

$$\mu(R) = \mu_0 + 1.086(R/a)^{1/4}$$

Results for the total luminosity, curve of growth, half-light radius and effective surface brightness of de Vaucouleurs profiles were given in equation (2.4.14–2.4.20). Dwarf ellipticals, on the other hand, have exponential profiles. The dividing line

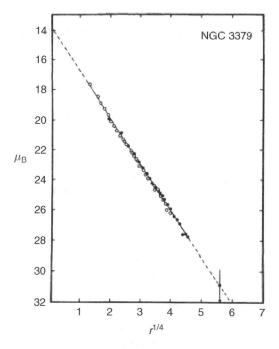

Figure 3.1 The surface brightness profile of the giant elliptical NGC 3379 which shows an excellent fit to an $R^{1/4}$ law (Reproduced with permission from de Vaucouleurs and Capaccioli, 1979).

between giant and dwarf early type galaxies appears to come at around $M_B = -18$ or $L_B \simeq 3 \times 10^9 L_\odot$. We will consider the dwarf ellipticals in more detail in Chapter 5.

In fact there may be a range of possible profile shapes when ellipticals are considered in detail. These can all be accomodated by generalising the de Vaucouleurs law to the Sérsic law,

$$I(R) = I_0 \exp\left(-(R/a)^{1/n}\right) \tag{3.2.1}$$

Here obviously $n=1$ corresponds to the exponential law and $n=4$ to the de Vaucouleurs law. Using the same notation as before, we can also write this as

$$I(R) = I_e \exp\left(-b\left((R/R_e)^{1/n} - 1\right)\right) \tag{3.2.2}$$

with $b \simeq 2n - 1/3$.

The Sérsic n parameters for ellipticals actually appear to fall into two groups. First, there are the really bright ellipticals that do have n close to 4, but then there are less bright Es for which n seems to decrease from 4 to about 2 as the luminosity decreases.

Although numerical simulations of the gravitational interactions between large numbers of stars do seem to lead to the de Vaucouleurs distribution from many possible starting points, there is no detailed theoretical justification for the $R^{1/4}$ law. Indeed we should remember that the surface brightness distribution we see is a projection onto the sky of the true 3-D distribution of stars, and in fact there is no analytic mathematical function for the density which projects exactly to the de Vaucouleurs form.

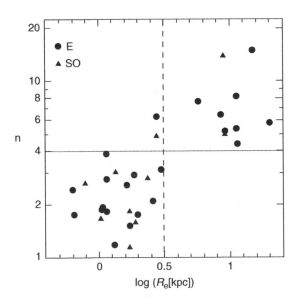

Figure 3.2 Variation of profile shape parameter n with elliptical galaxy size (Reproduced with permission from Caon, Capaccioli and D'Onofrio, 1993).

If we start with a simple form of the density, say a density that varies as a power law in true 3-D radius, $\rho \propto r^{-\gamma}$, we can see that the surface density or surface brightness as a function of projected radius on the sky will be

$$I(R) \propto \int_0^\infty (h^2 + R^2)^{-\gamma/2} \, dh \tag{3.2.3}$$

where h is measured along the line of sight through the galaxy (Figure 3.3). If we now make a change of variables to $g = h/R$, the integral becomes

$$\int_0^\infty \frac{R \, dg}{R^\gamma (g^2 + 1)^{\gamma/2}} = R^{-\gamma+1} G(\gamma) \tag{3.2.4}$$

where

$$G(\gamma) = \int_0^\infty (g^2 + 1)^{-\gamma/2} \, dg \tag{3.2.5}$$

is a number which depends only on γ. Thus a power law volume density of slope $-\gamma$ projects to a slope of $-\gamma + 1$ in surface density, so conversely an observed profile of slope $-\delta$, say, must be the projection of a 3-D density distribution $r^{-\delta-1}$. Note that in general we will use r for 3-D separations and R for projected radii on the sky, but in any case it should be clear from the context which is being referred to.

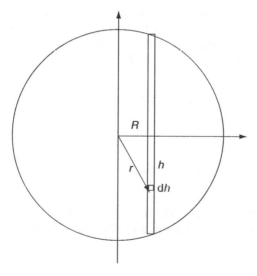

Figure 3.3 Schematic diagram illustrating the projection of volume density to surface brightness.

In fact, there are some simple forms which give reasonable approximations to elliptical galaxy profiles. Reynolds, back in 1913, and later Hubble, suggested

$$I(R) = I_0(1 + R/R_0)^{-2} \tag{3.2.6}$$

which obviously approximates a power law of slope -2 in the outer parts, but flattens towards the centre.

Similarly there are some useful approximations to a suitable density function; for instance, the Hernquist density

$$\rho(r) = \frac{M}{2\pi} \frac{r_c}{r(r_c + r)^3} \tag{3.2.7}$$

Calculating the same integral as before, we obtain a projection which is very close to the de Vaucouleurs law over a considerable range of radii.

The largest elliptical galaxies are the cDs which dominate clusters.[1] These have profiles which follow the $R^{1/4}$ law for perhaps 30 kpc, but then exhibit an excess of light above the continuation of the standard profile. This envelope, which can contain a substantial fraction of the total luminosity and extend to 100 kpc or more, is conjectured to form from material captured from other cluster galaxies or from the general intracluster medium.

As well as being related to the profile shape, the luminosity of an elliptical galaxy correlates tightly with the scale length of the profile. Since the total luminosity L is proportional to $I_0 a^2$ (or equivalently to $I_e R_e^2$), this also implies a correlation between luminosity and surface brightness. The correlation goes in the sense that the brightest Es are also the largest. However, the increase in size more than accounts for the increase in luminosity, so the surface brightness actually goes down. This 'Kormendy relation' is illustrated in Figure 3.4, which again shows that dwarf ellipticals are quite different to atleast some of their giant brethren, as their luminosity–surface brightness relation goes in the opposite direction. A good fit to Kormendy's original relation for giant Es is (for our value of H_0)

$$\mu_e = 20.2 + 3.0 \log R_e \tag{3.2.8}$$

or the equivalent

$$M_B = -19.3 - 2.0 \log R_e \tag{3.2.9}$$

Note that these imply $L \propto R_e^{0.8}$ and $I_e \propto R_e^{-1.2} \propto L^{-1.5}$. We will return to these relations when we consider the dynamics of elliptical galaxies in sections 3.8 and 3.9.

---

[1] Though this is not the origin of the name; in fact the 'c' derives from the stellar classification of supergiant stars and the 'D' from diffuse.

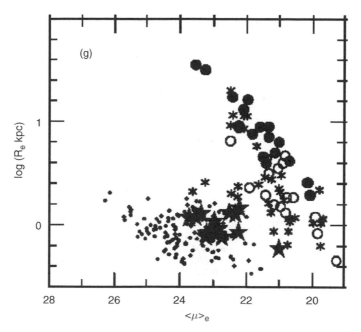

Figure 3.4 The Kormendy relation between surface brightness and scale size and for both giant (symbols) and dwarf ellipticals (dots) (from Graham and Guzman, 2003).

Before leaving the observed profiles, we should look at what happens right in the centres of elliptical galaxies. This really requires observations from space since the 'seeing' at a ground-based telescope will blur out detail in the central $1''$ or so. Since a typical elliptical has an effective radius of a few kpc, the scale length a is only around 1 pc. Thus the region where the intensity drops by the first few factors of e in a de Vaucouleurs profile will subtend a tiny angle even for quite local galaxies. With the HST, though, its resolution of $0.05''$ makes the exploration of the central regions feasible. Very luminous ellipticals are generally found to have a 'core', that is a region where the intensity is nearly constant. However, in less luminous ellipticals the surface brightness appears to continue rising steeply right in to a 'cusp' at the centre. The so-called 'nuker' law, introduced to cover these possibilities, is a double power law form

$$I(R) \propto \left(\frac{R_b}{R}\right)^{\gamma} \left[1 + \left(\frac{R}{R_b}\right)^{\alpha}\right]^{\frac{\gamma - \beta}{\alpha}} \tag{3.2.10}$$

Recall that for an intensity profile approximating a power law $I \propto R^{-\delta}$, the deprojected 3-D density must vary as $\rho \propto r^{-1-\delta}$; so for a cuspy profile, the actual density must be rising faster than r^{-1}. The luminosity density right at the centre of cuspy ellipticals can be of order $10^6 L_\odot\, pc^{-3}$.

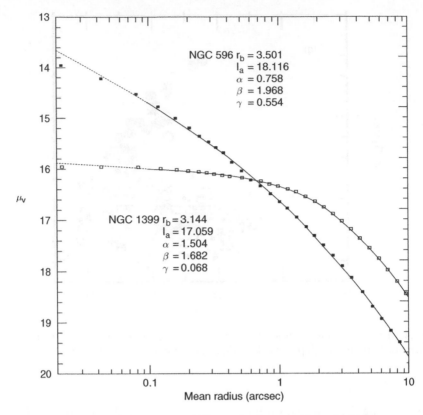

Figure 3.5   HST profiles of the centres of elliptical galaxies with and without a central core (Reproduced with permission from Lauer *et al.*, 1995).

3.3   Shapes

So far, despite their evident non-sphericity (a spherically symmetric density distribution obviously cannot project to an elliptical image on the sky), we have not considered the true 3-D shape of ellipticals. If we assume that they have regular isodensity surfaces, then there are three possibilities. They could be oblate spheroids, squashed down at the poles like the Earth, or prolate spheroids, with one long and two short axes like a rugby ball, or they could have all three principal axes of different lengths, a tri-axial ellipsoid (or squashed rugby ball!). At first sight the oblate case would seem the most likely, with the galaxies flattened by rotation in the same way as fluid bodies like the Sun. Note that an oblate spheroid clearly always looks less (or at most equally) flattened in 2-D projection than it is in 3-D. Thus we will need a range of true flattenings at least up to the 1:0.3 axis ratio seen for E7 galaxies. In fact one can see that this is true for prolate spheroids, too. From the majority of viewing angles, oblate spheroids should appear nearly circular, so the

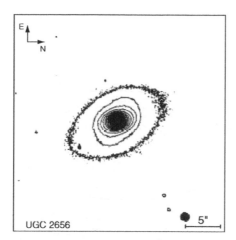

Figure 3.6 A galaxy with twisting isophotes (Reproduced with permission from Nieto *et al.*, 1992).

relatively small number of E0 galaxies suggests that ellipticals are not (or at least not usually) oblate.

In fact, it seems likely that most luminous ellipticals are tri-axial. Though it is difficult to picture (at least, the author finds it so!), the projection of general tri-axial isodensity surfaces in 3-D (even if all aligned with the same axes) can give rise, for many viewing angles, to elliptical isophotes (i.e. contours of equal surface brightness) which are *not* aligned.

Even for a tri-axial spheroid with twisting isophotes, the isophotes themselves are still perfect ellipses. However, not all elliptical galaxies are exactly elliptical! Two characteristic deviations are seen, either 'boxy' or 'disky' isophotes. Disky isophotes are more elongated along the major axis (so pointier) than a true ellipse, whereas boxy isophotes are compressed slightly along the axes, making them look more rectangular.

These distortions can be quantified by expressing the deviation of the shape of the isophotes from perfect ellipses as a Fourier expansion using polar coordinates. Thus we write the distance of an isophote from the galaxy centre, relative to the best-fitting ellipse (with semi-major and semi-minor axes a and b), as a function of the so-called ellipse parameter t, as

$$\delta R(\theta) = \sum (a_n \cos nt + b_n \sin nt) \qquad (3.3.1)$$

where t is related to the polar angle θ by $a \tan \theta = b \tan t$, and for simplicity we have assumed the major axis to be along $\theta = 0$.

The $n = 0$–2 terms just represent the best-fitting perfect ellipse, while in practice the a_3, b_3 and b_4 terms are all negligible. The isophote distortions are therefore measured by a_4. If it is positive, the isophote is extended along the lines where $4t = 0$, 2π, etc,

(a) Boxy isophotes

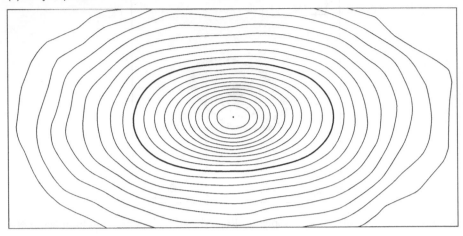

(b) Disky isophotes

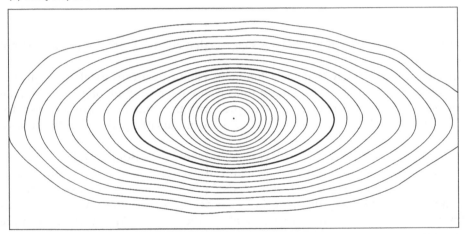

Figure 3.7   (a) Boxy (top) and (b) disky (bottom) isophotes in ellipticals (from simulations Reproduced with permission from Naab, Burkert and Hernquist, 1999).

or $\theta = 0$, $\pi/2$, π and $3\pi/2$, that is it is disky. If a_4 is negative, the isophote is shortened along these directions, making it boxy.

Other properties of the galaxies tend to correlate with the isophote shapes. For instance, very luminous ellipticals are usually boxy. Boxy ellipticals are also more likely to be radio and X-ray emitters. Medium-luminosity ellipticals are more likely to be disky, and also to be oblate and faster-rotating (p. 69). Disky ellipticals can be seen as obvious intermediates between the other ellipticals and S0 galaxies. S0s themselves are, in some senses, neither one thing nor the other. They clearly resemble ellipticals in that they are seen purely as smooth ensembles of stars, with no detailed small-scale structure and appear to be a smooth continuation of the disky ellipticals,

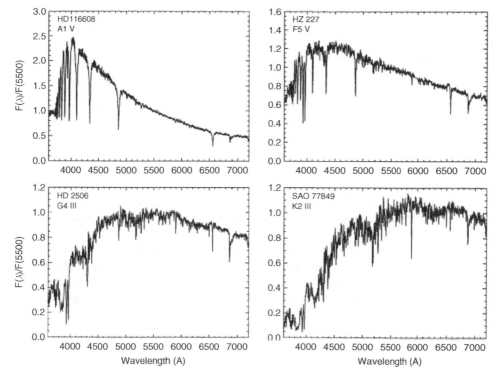

Figure 3.8 Spectra of stars of different types (adapted from Jacoby, Hunter and Christian, 1984).

but now the disc contribution can be large. Typically, the bulge contributes about 50–60% of the light, but the disc-to-bulge ratios can vary from about 0.1 to 2.

3.4 Stellar populations

So far we have said relatively little about the stars which make up galaxies. No giant ellipticals are sufficiently close to us that we can see individually any but the very brightest stars in them. Nonetheless, we can still infer what their population of stars must be like.

To do this we must examine the detailed spectrum of elliptical galaxies. We know that there is no sign of recent star formation activity in the form of bright blue stars or glowing gas in HII regions, as seen in spirals and irregulars, and this is confirmed by the lack of emission lines from hot gas in elliptical galaxy spectra. This was known already to the pioneer spectroscopists in the late 19th century and they also soon discovered the similarity of the spectra of elliptical nebulae to those of fairly cool stars.

Stars are traditionally divided into spectral types, based primarily on the strength of the absorption lines of various elements, caused by absorption of photons in the cooler outer layers of the stars. In particular, the stars with the strongest Balmer lines of hydrogen were assigned to spectral type A and those with the next strongest to type B. However, a more useful sequence for most purposes is based on the stars' colours or surface temperatures. Thus main sequence stars, burning hydrogen to helium in their cores, can be ordered in terms of decreasing temperature as types O, B, A, F, G, K and M. The main sequence is in fact a mass sequence and runs across the Hertzsprung–Russell diagram (section 1.6) from the hottest blue, high luminous O and B stars, which are the most massive, down to low-mass, cool, red M stars at low luminosities. The latter are often called red dwarfs; the even dimmer 'brown dwarfs' have been given the type letter L. Each type is subdivided into numerical classes 0–9, so that we get types like G2 and M5. The general change in the shape of the spectra as we run through the spectral types roughly matches that of black body spectra with temperatures ranging from above 30 000 K for O stars, through 6000 K or so for G stars like the Sun, down to 3000 K or less for M stars.

In terms of the spectral types, again working from O to M, we see first the Balmer lines of neutral hydrogen. These are not particularly strong in O stars as the hydrogen is almost all ionised, but increase to a maximum for A stars. The Balmer lines occur due to electrons in the first excited state absorbing photons of specific energies in order to jump to higher energy states of the hydrogen atom. Any photons which are energetic enough, though (i.e. of short enough wavelength), can remove the electron entirely (i.e., ionise the atom). Thus any photon with a wavelength short-ward of the 'Balmer Limit' at about 360 nm can be absorbed and the spectrum drops off sharply at this point.

As we reduce the temperature further, the Balmer lines get weaker again, but the absorption lines due to other elements become more prominent. In G stars the most important are the two Sodium D lines in the yellow, close to 590 nm (the wavelengths emitted by sodium street lights), the Magnesium b feature at about 520 nm, the so-called G band of the CH radical at around 430 nm and the H and K lines of singly ionised calcium (denoted CaII)[2] at 400 nm. In reading the astronomical literature, one should be aware that optical astronomers more usually use Ångstroms (Å) rather than nm, where $10\,\text{Å} = 1\,\text{nm}$. The combined effect of the Balmer limit and the CaII H and K lines in sharply cutting off the spectrum is therefore known as the '4000 Å break'. It is a key feature of the spectra of all intermediate temperature stars. In even cooler stars, molecules such as titanium oxide (TiO) provide the most characteristic absorption bands at the red end of the spectrum.

Now if we consider a galaxy as the sum of many stars, then the overall spectral shape and any features seen will tell us about the mix of stars the galaxy must contain. The hottest stars in the galaxy will dominate the blue end of the spectrum, and the lines we see there should be characteristic of O, B or A stars if they are present. Likewise the red light from a galaxy should be produced mostly by its K and

---

[2] In this notation, OI, for example, represents neutral oxygen, OII is singly ionised oxygen (i.e. the positive ion O^+), O III is twice ionised, and so on.

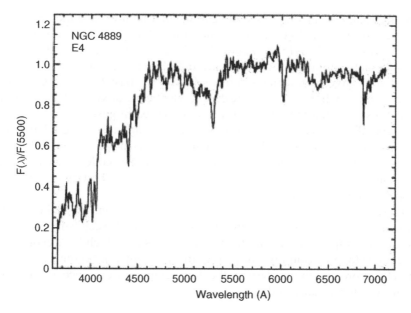

Figure 3.9   The characteristic spectrum of an elliptical galaxy (Reproduced with permission from Kennicutt, 1992).

M stars (assuming there are enough of them), so we should see the features typical of these at the long-wavelength end of the spectrum.

3.4.1   Stellar lifetimes

The amount of light a main-sequence star generates is related to its mass M (through the equations of stellar structure and nuclear energy generation) approximately as[3]

$$L \propto M^\alpha \qquad (3.4.1)$$

with $\alpha \simeq 3$ for masses below about $0.5M_\odot$ and $\alpha \simeq 4$ above that. Above about $10L_\odot$ the slope flattens out again to $\alpha \simeq 2$. Thus massive stars generate more light per unit mass than do smaller stars. Typical O and B stars – an O5 star has mass around $40M_\odot$ and a B4 star about $5M_\odot$ – have luminosities in the range $\sim 10^6$ down to $100L_\odot$.

A corollary to this is that massive stars must burn their available fuel faster than low-mass ones. Assuming that a more or less fixed fraction ($\sim 10\%$) of a star's original hydrogen content is burned to helium at the core, the time it can remain on the main sequence will be

$$\tau_{\mathrm{ms}} \propto \frac{M}{L} \propto M^{1-\alpha} \qquad (3.4.2)$$

---

[3] We will not go into the details of stellar structure here. The interested reader should consult one of the many text books in this area.

Using the values of α from above, we can see that high-mass stars have very short lifetimes relative to stars like the Sun. The Sun will stay on the main sequence for about 10^{10} years, while a $15M_\odot$ star will exhaust its fuel in only 10^7 years. Thus seeing a substantial contribution from luminous blue stars in a galaxy's spectrum will imply that stars have been formed in the galaxy in the past 10^7 years. Conversely, if we see no sign of such stars – as is the case in elliptical galaxy spectra – then we can infer that no star formation has taken place in the recent past.

More generally, the shape of a galaxy's spectrum can give us the relative number of stars of different types, and thus masses. This is nowadays usually done by running a 'population synthesis' model. This is a computer code which adds together different fractions of each type of stellar spectrum until the best match to the actual galaxy spectrum is obtained.

Of course, not all stars are on the main sequence. From the point of view of generating significant amounts of light, it is the giants, and specifically the red giants, that are of most importance in ellipticals. These are stars which have evolved off the main sequence (due to running out of fuel in their centres), becoming much larger and thereby more luminous despite the low temperatures, since via the Stefan-Boltzmann law for a black body,

$$L = 4\pi r^2 \sigma T^4 \tag{3.4.3}$$

Low-mass stars evolve roughly vertically in the H–R diagram, while more massive ones cool down and evolve 'sideways' across the plot to join the 'giant branch' (Figure 3.10), with hydrogen burning in a shell outside the core before the helium in the centre starts to be processed into heavier elements.

3.4.2   Stellar population evolution

By following the same sort of reasoning as above for the absence of recent star formation, we can now deduce how long ago star formation ceased. This is easier to picture in, say, a globular cluster in our own Galaxy where we can see the individual stars, but the same principle will apply to the integrated spectrum of a galaxy. Consider, then, forming a single generation of stars with some distribution of masses. This distribution is called the 'initial mass function' (IMF). These stars will populate the main sequence, but after a short time the most massive (bluest) stars will move off the main sequence to become giants. The main sequence will thus become truncated at the bright end. As time goes on, stars of lower mass will also evolve into red giants and the 'main sequence turn-off' will move gradually downwards. This can best be seen by comparing the H–R diagrams of star clusters of different ages, as in Figure 3.11. The colour of the overall stellar population will change with time, becoming redder as we replace blue main-sequence stars by red giants, and the red colours of ellipticals have long been used to deduce that they (like GCs) have very old stellar populations, with the stars having formed some 10^{10} years ago. The strengths of absorption features in the spectra, particularly the Balmer lines, are also an age indicator.

In addition, it will be evident that in stellar populations with no recent star formation it is the red giants which actually dominate the total light (since they are

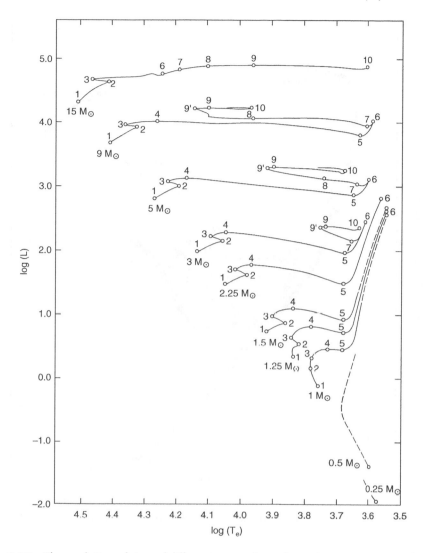

Figure 3.10 The evolution of stars of different masses from the main sequence onto the giant branch in the H–R diagram (Reproduced with permission from Iben, 1967).

hugely more luminous than red dwarfs). We can then make a simplified, but very useful, calculation of how the total luminosity of an elliptical galaxy should change with time. Essentially, the luminosity will be proportional to the number of red giants present at any time, t, after the stars formed. This will be the same as the number of stars with main sequence lifetimes between $t - \Delta t_g$ and t, where Δt_g is the time that stars remain as red giants. Notice that since main sequence lifetimes are functions of mass, we can also write this the other way round, giving the mass of a star which has a main-sequence lifetime t_{ms}, as $M(t_{ms})$. Thus the given range of ages will also correspond to a range of masses, $M(t_{ms})$ to $M(t_{ms} - \Delta t_g) = M(t_{ms}) + \Delta M$.

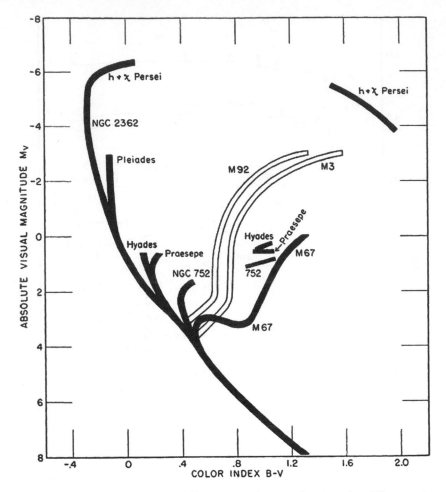

Figure 3.11 The H–R diagrams of different age clusters, showing the difference in main sequence turn-off points (Reproduced with permission from Johnson and Sandage, 1955).

So, finally, if we write the IMF as the number dN of stars formed with masses between M and $M + dM$, then the number of red giants at time t will be

$$N_g = \frac{dN}{dM} \times \Delta M = \frac{dN}{dM}\frac{dM}{dt_{ms}}\Delta t_g \qquad (3.4.4)$$

If we have a power law IMF, with $dN/dM \propto M^{-1-x}$, and recall that the main sequence lifetime $t_{ms} \propto M^{-1/\theta}$ with $\theta \simeq 1/3 (= 1/(\alpha - 1)$ in our earlier notation), then we obtain

$$N_g \propto t^{-1+\theta x} \qquad (3.4.4)$$

Assuming a standard 'Salpeter' IMF as supported observationally, with $x \simeq 1.35$, and taking the luminosity as proportional to the number of red giants we then end up with

$$L \propto t^{-1+\theta x} \simeq t^{-0.6} \qquad (3.4.5)$$

This kind of calculation, pioneered by Jim Gunn and Beatrice Tinsley in the mid-1970s, gives us our first insight into how a whole galaxy should evolve over cosmic time. In particular, it shows that the luminosity of elliptical galaxies should have been higher in the past. We would expect to see this effect if we study galaxies at high redshift, because of the long light-travel or 'look-back' time involved. Of course, one might argue (correctly) that stars of different masses lead to red giants of different luminosities and different lifetimes. However, these effects nearly cancel out; as on the main sequence, more massive giants are brighter but shorter-lived. More detailed calculations of elliptical galaxy evolution – made in the same way as we discussed for population synthesis models, but now including the known evolution of each mass of star – give quite similar results to our approximate one.

3.4.3   Surface brightness fluctuations

Before leaving the topic of stellar populations, we can note an interesting application to the distance determination problem. If we again consider red giants to dominate the luminosity, then we can equate the surface brightness of a galaxy to the number density of such stars. Suppose we have a fixed angular-sized aperture on the sky (perhaps a detector pixel), then for two otherwise identical galaxies whose distance differs by a factor f, there will be f^2 more stars in the aperture for the distant galaxy, each of which is $1/f^2$ times as bright, preserving the surface brightness. But the Poisson ($\sqrt{N}$) noise in the surface brightness will be *smaller* by a factor $\sqrt{f^2}$, i.e. f, because of the higher number of stars contributing. Thus the more distant galaxy will look 'smoother', and the level of 'surface brightness fluctuations' (SBFs) will therefore provide a distance estimate. An absolute calibration can come either from galaxies of known distance or, more usually, from a direct calculation of what the SBF should be for a given population of stars (so that the simplification to equal luminosity red giants is not actually needed in the end).

3.5   Metallicity

Another aspect of stars in relation to galaxy evolution, that we have not yet mentioned, is their chemical composition. Clearly the nuclear processing in stars leads to the creation of heavier elements (known collectively in astrophysics as 'metals') from hydrogen and helium. By convention, the fraction by mass of hydrogen in a star is written as X, that of helium as Y and that of metals (i.e. everything else) as Z. For normalisation we can note that about 2% of the Sun is made up of heavy elements, so $Z_\odot = 0.02$. Given that old stars shed their outer layers, either quietly via stellar winds and the production of planetary nebulae or spectacularly in supernova explosions, much of this processed material will be returned to the ISM. It can then be incorporated into the next generation of star formation from interstellar clouds. We thus expect that as a galaxy evolves it will produce stars of higher and higher metallicity. Since the atoms or ions of these metals can absorb photons (as in, for example, the Mg and Ca lines we met earlier), the more the heavy elements are present, the more light will be absorbed. As this happens to occur preferentially at the blue end of the optical spectrum, this loss of blue light will make stars with higher metallicity look redder than their metal-poor counterparts. A galaxy made up of metal-rich stars will similarly look redder than one composed of metal-poor stars. Note that this effect is the same as looking at an older population of stars, so if we have only the optical colour of a galaxy to go on, this will lead to the so-called 'age–metallicity degeneracy' in early type galaxies; that is, a galaxy which looks bluer than others might be younger *or* more metal-poor.

This leads us on to the question of the overall colours of early type galaxies. We have concluded already that they will be red, at least compared to spirals, but the points made above might lead us to doubt that they will all be equally red. In fact, if we plot the colour, say the optical $U–V$ or the optical to near infrared $V–K$, against the absolute magnitude, then we find a very tight correlation. This is seen particularly well if we plot this diagram for all the early type galaxies in a cluster like Coma or Virgo. Less-luminous early types are slightly bluer than brighter ones. Of the two possibilities, of their stars being young or metal-poor, it turns out that detailed spectroscopic analysis supports the latter (Figure 3.13). This is taken to imply a correlation between the metallicity of an elliptical galaxy and its mass. A lower-mass galaxy will have a lower escape velocity, so it will be easier for heavy elements produced in one generation of stars to be lost from the galaxy, via 'galactic winds',

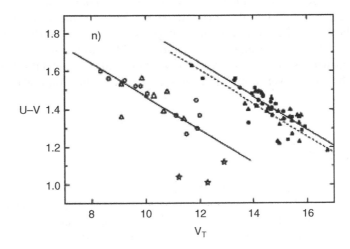

Figure 3.12 The colour–magnitude relation for early type galaxies in the Coma and Virgo Clusters (Reproduced with permission from Bower, Lucey and Ellis, 1992). Note that there is a shift between the data for the two clusters as they are at different distances (the axis here is apparent magnitude).

before they can be incorporated into later generations of star formation, thus keeping the overall metallicity down. The correlation also carries on from giant ellipticals down to dwarf ellipticals, while Local Group dwarf spheroidals have even lower metallicities again (Chapter 5). Of course, not all stars in a galaxy need be exactly the same metallicity, and even in ellipticals the stars at the centre tend to be more metal-rich than those further out. Indeed, the stars in the central parts of giant ellipticals are probably the most metal-rich found anywhere, with metallicity about twice that of the Sun, i.e. $Z \simeq 2Z_\odot$.

Note that in Figure 3.13 we have used another form of notation for metallicity, the strength of the magnesium 'index', a measure of the prominence of the magnesium feature at around 517 nm in a galaxy's spectrum. Other ways of denoting metallicity will also be encountered in the astronomical literature. These are generally in terms of the number of atoms of a given element relative to hydrogen. For example, $12 + \log(O/H)$ is just the logarithm of the number of oxygen atoms for every 10^{12} hydrogen atoms. For the Sun, $12 + \log(O/H) = 8.93$. In another convention, [Fe/H] means the logarithm of the ratio of Fe to H atoms in an object, compared to what it is in the Sun; generally, if N_X represents the number of atoms of element X, then $[X/H] = \log(N_X/N_H) - \log(N_{X\odot}/N_{H\odot})$.

The tightness of the colour–magnitude correlation also carries important information, as it implies that all elliptical galaxies must be at essentially the same stage in their evolution. That is, they are all made of stars of the same age. This has usually been interpreted as meaning that elliptical galaxy formation was 'coeval', a once only event in the history of the universe. Strictly though, it says that all the *stars* in giant ellipticals formed at the same time. They might have been assembled into the galaxies we see at a later stage.

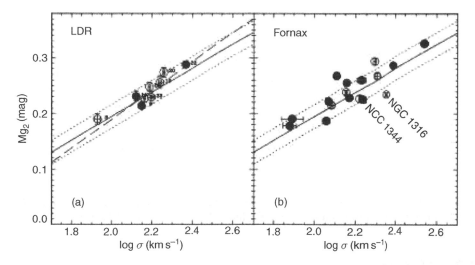

Figure 3.13 The metallicity–luminosity relation for two samples of elliptical galaxies, plotted as a correlation between the strength of the magnesium feature in the galaxy's spectrum and its velocity dispersion (Reproduced with permission from Kuntschner *et al.*, 2002).

3.6   Globular clusters

Before leaving the stellar content of early type galaxies, we should mention their GCs. Elliptical galaxies have systems of GCs very much like those around our own Galaxy. Indeed, the distribution of luminosities of globulars (i.e. the Globular Cluster Luminosity Function, or GCLF), between $\sim 10^4$ and $\sim 10^6 L_\odot$, seems to be virtually identical, irrespective of the galaxy they surround. Note that this apparent universality of the GCLF shape makes the corresponding characteristic V magnitude, $M_{GC} \simeq -8.5$, where the function turns over into a useful standard candle and thereby distance indicator.

On the other hand, the number of GCs around a galaxy can vary widely. Our Galaxy has some 130 or so, while NGC 1399, the first-ranked (i.e. brightest) galaxy in the Fornax Cluster, has around 1600. As one might expect, the numbers go up as the luminosity of the galaxy goes up, but we can define a 'specific frequency' as the number of globular clusters per unit galaxy light (usually per $10^8 L_\odot$). Most spiral galaxies have similar values of this quantity, $S_N \simeq 0.4$, but ellipticals generally have larger specific frequencies, around 2. The largest ellipticals in clusters, like NGC 1399 or M87, have even larger retinues of GCs for their luminosity.

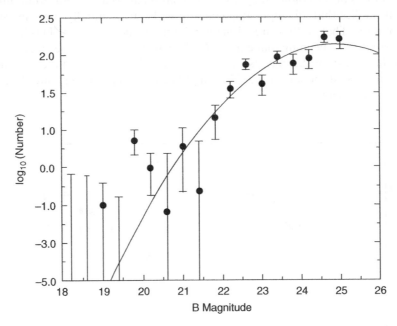

Figure 3.14   The luminosity function of globular clusters using data for clusters around NGC 1399 (Reproduced with permission from Bridges, Hanes and Harris, 1991).

3.7   Hot gas

As we have seen, stars lose mass as they evolve. In our Galaxy and other spirals this will mix in with the existing ISM, providing fuel for further star formation. The same must have happened in the first stages of evolution of early type galaxies in order to accommodate the increase in metallicity up to the observed levels. Currently, most early types contain negligible amounts of the neutral hydrogen which dominates the ISM in spirals; so where has the mass lost, since star formation ceased, gone to? For a standard IMF the stars in an elliptical of current luminosity $\simeq 10^{10} L_\odot$ will collectively lose 1 or $2 M_\odot$ each year, so over several billion years something like $10^{10} M_\odot$ of gas should be cycled back into the ISM.

In fact, after many astronomer-years of effort trying to solve this puzzle, it turned out that the gas had not gone anywhere, it was merely in a then unobservable form, that is extremely hot. This component was finally observed when X-ray satellite observatories were launched and giant ellipticals were seen to be prominent X-ray sources.

In retrospect, this should have been an obvious repository for the gas if we assume that the kinetic energy per unit mass of the material shed from stars is roughly the same as that of the stars themselves as they move around under their mutual gravitational attraction. We know from ordinary thermodynamics that the energy of gas particles is just a measure of temperature

$$\frac{3}{2} k_B T = \frac{1}{2} m <v^2> \tag{3.7.1}$$

where k_B is Boltzmann's constant, m is the average particle mass and $<v^2>$ is the mean square speed of the stars. Since very hot gas will be completely ionised, the mean particle mass will be half the proton mass (the electrons having negligible mass compared to the protons). Thus

$$T = \frac{m_p <v^2>}{6k_B} \simeq 2 \times 10^6 \frac{<v^2>}{\left(500 \, \mathrm{km \, s^{-1}}\right)^2} \mathrm{K} \tag{3.7.2}$$

Here the velocities are normalised by a typical value for a moderately large elliptical galaxy. Galaxies like M87 actually show an emission spectrum of gas at about 3×10^7 K.

Above 10^7 K, recombinations of electrons onto atomic nuclei are extremely rare, so it is hard for the gas to cool via electrons dropping to lower energy levels by the emission of a photon in an emission line. Instead we have mainly 'free-free', or 'thermal bremsstrahlung', radiation from electrons scattering off (but not being captured by) protons.

Bremsstrahlung (German for 'braking radiation') depends on the frequency of collisions between protons and electrons and hence on the number density of protons multiplied by that of electrons. But both of these are essentially equal to the hydrogen density n_H so

$$L_X \propto n_H^2 \Lambda(T) \tag{3.7.3}$$

where Λ is a measure of the cooling rate for gas at temperature T. The 'cooling time' that would be required to radiate away all the energy of the gas is then

$$t_{\mathrm{cool}} = \frac{3}{2} \frac{(n_e + n_p) k_B T}{n_e n_p \Lambda(T)} \simeq \frac{3 k_B T}{n_H \Lambda(T)} \tag{3.7.4}$$

For bremsstrahlung $\Lambda(T) \propto T^{1/2}$ so hotter gas takes longer to cool. At typical particle densities less than $10^{-8}\,\mathrm{m}^{-3}$, $t_{\mathrm{cool}} > 1\,\mathrm{Gyr}$.

3.8 Dynamics

Summing up some points from the previous section, if we assume that elliptical galaxies contain only old stars and that the IMF is the same for all galaxies, it is clear that (at any given epoch) the stellar mass of a galaxy determines its luminosity. Further, we have seen that luminosity correlates with colour, and hence metallicity. Earlier we noted that luminosity correlates with surface brightness and/or size- and intensity-profile shape and probably also affects more detailed features such as cusps/ cores and disky/boxy isophotes. This might then lead us to enquire whether, like stars on the main sequence, elliptical galaxies are effectively a 'one-parameter family', with all their characteristics fixed (at least to a good approximation) just by the mass. To look into this further we need to explore the dynamics of elliptical galaxies.

Most elliptical galaxies appear flattened. The obvious interpretation of this was as an effect of rotation, as with the flattening of the Earth or the Sun. Unfortunately, when technological advances allowed the rotation to be measured, the obvious turned out to be wrong!

The actual measurements almost always have to be made by measuring the absorption lines in the spectrum of a galaxy. If we can spatially resolve the spectrum, that is determine the spectrum at different points across the galaxy, then we may be able to see a differential Doppler shift from one side of the galaxy to the other. Such observations are traditionally done by using a 'long slit' on the spectrograph, so that we simultaneously take in light from a slice along (usually) the major axis of the elliptical image.

This results in a 2-D image on the spectrograph detector, where one dimension – the dispersion direction – shows the different wavelengths, and the other – the spatial direction – represents positions along the slit. Nowadays, this may be replaced by an 'integral field unit' which measures the spectrum at individual positions in the galaxy by using a number of separate optical fibres to feed the spectrograph. Whichever method is used, the signature of a rotating galaxy is the presence of tilted spectral lines, indicating the changing Doppler velocities with position.

Measurements of this sort for giant elliptical galaxies were first made in the late 1970s and immediately showed them to be much more slowly rotating than spirals, indeed too slowly rotating to cause significant flattening. To see what does give rise to their shapes, first consider what could hold the stars up against their own self-gravity in a completely non-rotating system. Clearly, just as for atoms in a gas cloud it must be their random kinetic motions (the 'temperature' of the system).

Now we know from ordinary dynamics that the faster a particle is fired outwards in a gravitational field, say from the Earth, the further it can go before being brought to a halt and falling back inwards again. In other words, by conservation of energy, the larger kinetic energy allows a larger change in the gravitational potential energy between the particle's starting point and its furthest excursion.

But if we consider the components of the particle's (i.e. star's) original outward velocity in the x-, y- and z-directions, then, roughly speaking, it will go furthest in the direction in which it has the greatest velocity. The same will be true if we consider three separate stars, each moving outwards along a principle axis with a different speed, and we can then extend this to a whole ensemble of stars. If the mean square velocity in the z-direction, $<v_z^2>$, is less than $<v_x^2>$ and $<v_y^2>$, for instance, then the stars will typically travel less far in the z-direction than the others. Thus our ball of stars will be less extended in the z-direction than in x- and y-directions and the system will look flattened. Different values for all the $<v_x^2>, <v_y^2>$ and $<v_z^2>$ will give a tri-axial shape. The same general considerations are still true if the mass is not all concentrated at the centre, as we have implicitly assumed.

The mean square velocities are usually referred to as velocity dispersions and written as σ^2. In elliptical galaxies the velocity dispersions are not usually determined from the spread of the velocities of individual stars, as they cannot be observed separately. Instead we observe the *width* of the spectral lines. For a given star, a spectral line such as Ca, H or K will be at a specific wavelength, but because all the stars are moving at different velocities the galaxy spectrum – the sum of those of all its constituent stars – will show a line broadened by an amount dependent on the velocity dispersion. Observationally, in order to remove any broadening effects due to the spectrograph (no instrument can measure with infinite resolution, of course), it is usual to also observe a single 'template' star of a suitable spectral type (typically a K giant for comparison with a giant elliptical). The required line in the star can be artificially broadened to match that in the galaxy by convolving the observed spectrum with a gaussian function. The width of the gaussian then gives the velocity dispersion in the galaxy. Notice that what we are actually measuring via Doppler shifts is the one-dimensional velocity dispersion radially along the line of sight, σ_r^2. It tells us nothing about the velocity dispersion in the two directions in the plane of the sky. Note too, that in general the velocity dispersion can vary with position within the galaxy. If it does not, we have an isothermal distribution (literally, the same temperature everywhere), and, in practice, galaxies are not too far from this ideal. Observationally, because of the concentration of light there, what we normally measure is the central velocity dispersion σ_0, though with more sophisticated techniques (and analysis) it is now possible to probe the variation of velocity dispersion with position in elliptical galaxies.

3.8.1   Rotation

Returning to the question of rotation, we should consider how much rotation we might require in order to generate a certain amount of flattening. Take a galaxy rotating in the x–y plane (i.e. about the z-axis) with rotational speed V. Let the velocity dispersion be isotropic, $\sigma_x^2 = \sigma_y^2 = \sigma_z^2$. Now, from the discussion above, we would expect that the distance the stars can go in the different directions should scale with the kinetic energies, K_x, K_y, K_z, associated with the respective velocity components. Thus, assuming for now that we view the system 'side-on', with y in the radial direction, the axis ratio should follow roughly

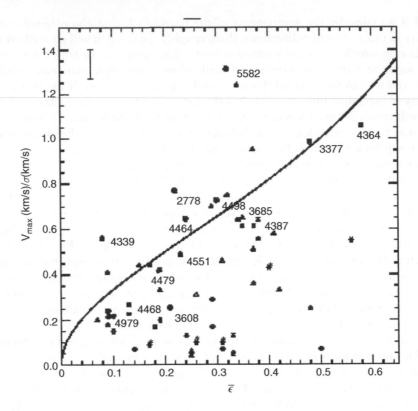

Figure 3.15 Plot of V/σ versus ellipticity ϵ. The numbered points are for low-luminosity galaxies, the stars and the triangles are for high- and medium-luminosity ellipticals respectively (Reproduced with permission from Halliday *et al.*, 2001).

$$\frac{b}{a} \simeq \frac{K_z}{K_x} \simeq \frac{\sigma_z^2}{V^2 + \sigma_x^2} \tag{3.8.1}$$

since in the plane there is a contribution from the ordered motion (rotation). Now with this orientation we get $\sigma_y^2 = \sigma_r^2$, where σ_r^2 is the observed radial velocity dispersion, and by isotropy this is also equal to the other components, so

$$\frac{b}{a} \simeq \frac{\sigma_r^2}{V^2 + \sigma_r^2} \tag{3.8.2}$$

Inverting this (and dropping the subscript r) gives us

$$\left(\frac{V}{\sigma}\right)^2 \simeq \left(\frac{a}{b} - 1\right) = \frac{\epsilon}{1 - \epsilon} \tag{3.8.3}$$

where $\epsilon = 1 - b/a$ is the observed ellipticity. A more sophisticated treatment, allowing properly for the projections of the velocities gives essentially the same result and moreover the equation applies even for inclined systems.

In order to get a flattening of, say, $b/a = 0.5$ (an E5), we would require $V/\sigma \simeq 1$. Even $b/a = 0.8$ implies $V/\sigma \simeq 0.5$. A typical giant elliptical has $\sigma \simeq 250 \text{ km s}^{-1}$ or more, so we require rotation speeds of the same order, close to those actually seen for spirals. Much lower rotation velocities are seen in most cases in practice, typically corresponding to $V/\sigma \sim 0.2$. Thus ellipticals cannot be oblate rotating ellipsoids with isotropic velocity dispersions, that is they must have the sort of anisotropic velocity dispersions discussed above.

In fact, as we can see from the observations summarised in Figure 3.15, some ellipticals *do* rotate fast enough to be rotationally flattened. These are generally the lower-luminosity ones and almost always have disky rather than boxy isophotes. They probably have 'buried' rotating discs within the main body of their stars, making them behave more like S0 galaxies.

3.8.2 The virial theorem

Having got this far, we should next consider the equilibrium of these systems a little more formally. Exactly as for a gas cloud or any other isolated dynamical system, the stars in a galaxy will satisfy the usual virial theorem

$$2K + U = 0 \tag{3.8.4}$$

providing the density distribution is not changing with time. The total kinetic energy for a collection of individual stars (each labelled by an index i) will clearly be

$$K = \sum_i m_i v_i^2 / 2 \tag{3.8.5}$$

If the stars' velocities are uncorrelated with their masses, then this reduces to

$$2K = M < v^2 > \tag{3.8.6}$$

where M is the total mass, and the mean square velocity is just what we have called σ^2.

The total gravitational potential energy is the sum of contributions from all *pairs* of particles (at positions r_i, r_j), so

$$U = \sum_{i>j} \frac{-G m_i m_j}{|r_i - r_j|} \tag{3.8.7}$$

where the condition $i > j$ prevents each pair being counted twice. Again assuming that there has been no separation of the stars by mass, it is clear dimensionally that this must be

$$U = -\frac{G M^2}{r_g} \tag{3.8.8}$$

where $1/r_g$ is some weighted average of the inverse separations of the stars, and hence must represent a characteristic extent of the system. For a spheroidal system we can then write this as

$$U = -\alpha \frac{GM^2}{R_e} \tag{3.8.9}$$

where R_e is our usual (projected) effective radius and α is some constant (for galaxies of a given radial density profile) of order unity.

Taking the limit of a large number of particles (stars) we can write the sum as an integral, so if we imagine building up the galaxy by successively adding thin shells of density $\rho(r)$ and mass $dM = 4\pi\rho(r)r^2 dr$ on top of the existing mass $M(r)$ inside radius r, then

$$U = -G \int_0^\infty \frac{M(r)dM}{r} \tag{3.8.10}$$

As a simple example, consider a uniform density sphere of radius r_s. The mass inside r is just $4\pi r^3 \rho/3$ and the shell mass is $4\pi r^2 \rho dr$, so we get

$$U = -G \int_0^{r_s} \frac{16\pi^2}{3} \rho^2 r^4 dr = -\frac{16}{15} \pi^2 G \rho^2 r_s^5 \tag{3.8.11}$$

But the total mass is $M = 4\pi r_s^3 \rho/3$ and we can calculate that the (projected) half-mass radius is $R_e = (1 - 2^{-2/3})^{1/2} r_s = 0.608 r_s$, so we have

$$U = -0.365 \frac{GM^2}{R_e} \tag{3.8.12}$$

which is evidently of the standard form with $\alpha \simeq 1/3$.

It is also evident from the foregoing that the mass of an elliptical galaxy should be related to its velocity dispersion via the virial theorem as

$$M\sigma^2 = \alpha \frac{GM^2}{R_e} \tag{3.8.13}$$

i.e.

$$M = \frac{1}{\alpha} \frac{R_e \sigma^2}{G} \tag{3.8.14}$$

Note that here σ^2 is the 3-D velocity dispersion, so for a near isotropic system $\sigma^2 \simeq 3\sigma_r^2$. Thus finally

$$M \simeq \frac{3\sigma_r^2 R_e}{\alpha G} \tag{3.8.15}$$

and if the mass-to-light ratio is M/L, then obviously the luminosity

$$L \simeq \frac{3\sigma_r^2 R_e}{\alpha G(M/L)} \tag{3.8.16}$$

Indeed, from dimensional considerations, it is clear that we must always have

$$L \propto \frac{\sigma_r^2 R_e}{(M/L)} \tag{3.8.17}$$

and recalling that also $L \propto I_e R_e^2$, we obtain

$$L \propto \frac{\sigma^2 (L/I_e)^{1/2}}{(M/L)}$$

giving finally

$$L \propto \frac{\sigma^4}{I_e (M/L)^2} \tag{3.8.18}$$

3.9   The Faber–Jackson relation and the fundamental plane

We should probably not be surprised then, if the luminosities of elliptical galaxies were correlated with their velocity dispersions and, indeed, it was noted in the early 1960s by Poveda and by Minkowski that this was the case. More modern work originates with relationship discussed by Sandy Faber and Robert Jackson in 1976. In its original form, the Faber–Jackson (F–J) relation follows the simplest prediction

$$L \propto \sigma^4 \tag{3.9.1}$$

or more quantitatively

$$\frac{L}{2 \times 10^{10} L_\odot} \simeq \left(\frac{\sigma_r}{200 \, \text{km s}^{-1}} \right)^4 \tag{3.9.2}$$

over the range roughly $50 \le \sigma_r \le 500 \, \text{km s}^{-1}$. Other estimates have suggested that the power of σ might be nearer to 3 than 4 and depend somewhat on the wavelength at which L is measured.

Recall that earlier we also had a relationship between an elliptical's surface brightness (or size) and its luminosity – the Kormendy relation. We can effectively combine this and the F–J relation if we examine the distribution of early type galaxies in the 3-D parameter space of L, I and σ. We then find that rather than occupying the whole of the space, the points for the galaxies all lie close to a plane, the so-called 'fundamental plane' or FP. Specifically, the three parameters are linked by a relation approximately

$$I_e^{5/3} R_e^2 \propto \sigma^{5/2} \tag{3.9.3}$$

or alternatively

$$R_e \propto \sigma^{5/4} I_e^{-5/6} \tag{3.9.4}$$

which we can also write in terms of the luminosity as

$$L \propto I_e^{-2/3} \sigma^{5/2} \tag{3.9.5}$$

for example.

We can think of the Kormendy relation (between L and I_e) and the Faber–Jackson relation (between L and σ) as projections of this tilted plane in 3-D onto the respective 2-D slices.

Notice, too, that as we have correlations between a galaxy's luminosity and one or more distance-independent observables (surface brightness and/or velocity dispersion), we can deduce L from measured values of I_e and σ. Then if we also measure the apparent luminosity, we can use the usual inverse square law to deduce the distance to the galaxy. Thus the F–J and FP relations provide very useful distance indicators in the range where we need to use the properties of whole galaxies to find distances.

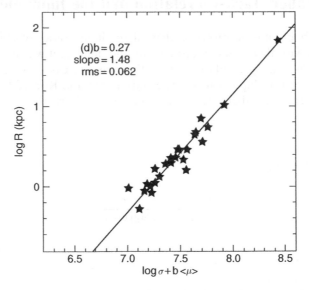

Figure 3.16 The Fundamental Plane for elliptical galaxies (Reproduced with permission from Graham and Colless, 1997).

3.9.1   Peculiar velocities

Such methods are often used in order to determine the 'peculiar velocities' of galaxies, that is their velocities relative to the overall Hubble expansion. If we know the distance D and the observed recession velocity cz then the peculiar velocity will be (to first order)

$$\nu_{\text{pec}} = cz - H_0 D \tag{3.9.6}$$

Note that, contrary to first appearances, this is not dependent on us getting H_0 correct, as this will cancel out with the assumed value of H_0 built into the luminosities used in obtaining distances from the F–J relation. Investigating velocity fields and galaxy 'flows' via these methods has become an important tool in the overall study of the large-scale distribution of mass in the universe, as it is the gravitational effect of the (fairly) local masses which will generate the peculiar motion of a given galaxy.

3.9.2 Mass-to-light ratios

Looking back at our virial theorem prediction we can see that, ignoring any variation due to different density distributions (called 'non-homology' and hidden in the parameter α in equation (3.8.9) and in the proportionality constant between L and $I_e R_e^2$), we must have

$$\frac{M}{L} \propto \frac{\sigma^2 R_e}{\sigma^{5/2} I_e^{-2/3}} \propto \sigma^{-1/2} R_e \left(\sigma^{5/2} R_e^{-2} \right)^{2/5}$$

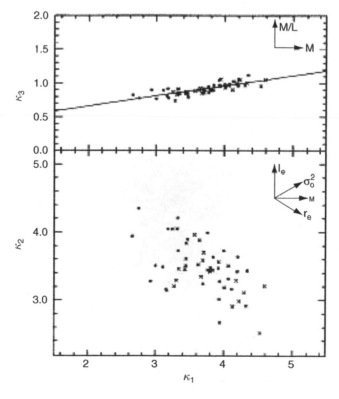

Figure 3.17 The k-space distribution for elliptical galaxies (Reproduced with permission from Bender, Burstein and Faber, 1994).

i.e.

$$M/L \propto \sigma^{1/2} R_e^{1/5} \tag{3.9.7}$$

As ellipticals lie close to a single plane in their original 3-D parameter space, we can choose combinations of parameters so as to make the plane 'level' as opposed to tilted. Thus in one projection we will see the plane exactly edge-on. For instance, writing $L = c_1 I_e R_e^2$ and $M = c_2 \sigma^2 R_e$, we can translate the physical coordinates into 'k-space' via

$$k_1 = \frac{1}{\sqrt{2}} \log \left(\frac{M}{c_2} \right) = \frac{1}{\sqrt{2}} \log \sigma^2 R_e \tag{3.9.8}$$

$$k_2 = \frac{1}{\sqrt{6}} \log \left(\frac{c_1}{c_2} \frac{M}{L} I_e^3 \right) = \frac{1}{\sqrt{6}} \log \left(\frac{\sigma^2 I_e^2}{R_e} \right) \tag{3.9.9}$$

$$k_3 = \frac{1}{\sqrt{3}} \log \left(\frac{c_1}{c_2} \frac{M}{L} \right) = \frac{1}{\sqrt{3}} \log \left(\frac{\sigma^2}{I_e R_e} \right) \tag{3.9.10}$$

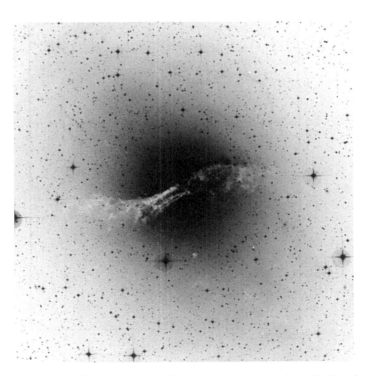

Figure 3.18   The dusty elliptical Cen A (UKST image courtesy of AAO). The obscuring dust lane appears pale in the usual (negative) image.

Observations show that with these variables we get the edge-on view in the $k_1 - k_3$ plane, viz.

$$k_3 = 0.15k_1 + \text{constant} \tag{3.9.11}$$

It is easy to check that this is close to the FP relationship given above.

In any of these representations it is evident that larger, more massive galaxies must have *higher* mass-to-light ratios than smaller ones (e.g. a large k_3 for a large k_1). This is consistent with our earlier observation that more luminous ellipticals are redder, since redder populations generate less light per unit mass.

3.10   Mergers

There remains, of course, the question of *why* ellipticals should follow the above (or indeed any) scaling relations. They must presumably arise as a result of either the way in which ellipticals form or the way in which they evolve (the 'nature versus nurture' argument). For the latter, since the relationships involve σ, this must be dynamical evolution, not just the evolution of the stellar population.

Although the stars in early type galaxies appear to be very old, it is likely that some ellipticals, at least, have only arrived at their current form quite recently. On the

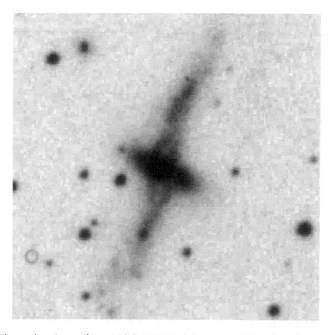

Figure 3.19   The polar ring galaxy NGC 4650A. (Courtesy: UK Schmidt Telescope.)

whole, ellipticals are notable only for their lack of features, but some, like the nearby example Cen A, show significant departures from the classical smooth elliptical appearance. Cen A itself displays a huge dust lane across its centre and in all about 5–10% of ellipticals show evidence for dust patches. This is, of course, also against the standard picture of early type galaxies lacking any significant cold interstellar medium, but in some cases we can directly observe the gas too. In a few we see a 'polar ring', where the gas loops over the 'top' of a flattened elliptical or over the bulge of an S0 at a large angle to the plane.

Even non-dusty ellipticals, when looked at in sufficient detail, can reveal interesting features. For instance, if we 'unsharp mask' the image of a bright elliptical by subtracting off a smoothed version of the image from the original (either by cunning photographic processing, as performed by David Malin for the original discovery of these features, or more prosaically these days by computer digital enhancement), we sometimes uncover 'shells'. These sharp-edged features lie inside the main body of light and often interleave, in the sense that partial shells alternate at either end of the major axis as their radii increase. In other systems one can find faint 'tails' and other non-axisymmetric features. More impressive tails can sometimes be seen in cases where a galaxy has a near neighbour, and some very bright ellipticals (such as brightest cluster members) show double nuclei.

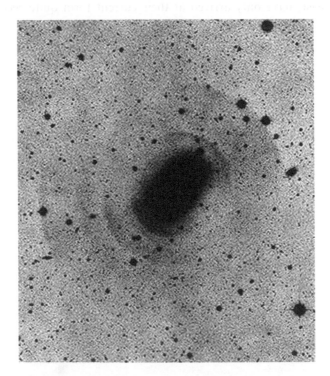

Figure 3.20 Reprocessed image of the galaxy NGC 3923, showing its shells. (Courtesy: David Malin and the Anglo-Australian Observatory.)

All these observations suggest that galaxies may interact or even merge with one another.[4] Indeed, the merging process can actually be seen taking place in systems like 'The Antennae', where the two main galaxies are in contact and tidal streams of debris, drawn out earlier in the encounter, are stretched out behind them. Later on, the two merging galaxies will look more like one irregular object and the tails will be less prominent. A different type of interaction, a headlong plunge of a smaller companion straight through the middle of a disc, can lead to a ring galaxy such as the famous Cartwheel.

If at least one of the galaxies is a gas-rich spiral or irregular, the interaction may funnel interstellar material down towards the centre of the merger product and drive a burst of star formation. The best-known nearby starburst galaxy is M82.

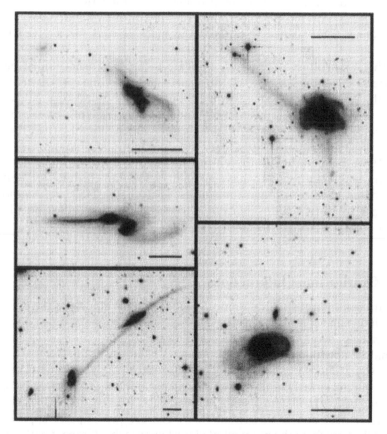

Figure 3.21 A montage of mergers believed to represent different stages in the process, with age increasing clockwise from bottom left (Reproduced with permission from Hibbard and van Gorkom, 1996).

---

[4] Galaxy interactions were first investigated by Erik Holmberg in 1940. He built a remarkable analogue 'simulation' with light intensity (also a $1/r^2$ law, of course) instead of gravitational force and using 37 light bulbs to represent each galaxy!

This contains some $2 \times 10^8 M_\odot$ of gas which, at its present star formation rate of a few $M_\odot$ per year, will be used up in $\simeq 10^8$ years. The most dramatic of the starburst galaxies are probably the ULIRGs – ultra-luminous infra-red galaxies[5] – which may have $10^9 - 10^{10} M_\odot$ of gas. Here the interstellar medium becomes so dense that most of the optical and ultra-violet emission from the star-forming region is absorbed by the surrounding thick dust clouds and reprocessed into $10^{12} L_\odot$ or more of FIR radiation.

Their star formation rates can be estimated to be

$$\dot{M}_* \simeq 5 \times 10^{-10} L_{60}/\epsilon \tag{3.10.1}$$

where ϵ ($\simeq 1$) is the fraction of the light re-emitted in the FIR and L_{60} is the 60-μm luminosity in $L_\odot$. This has an uncertainty of a factor of two either way, depending on the assumed IMF (since this alters the mass needed for a given luminous output).

One other by-product of mergers appears to be the formation of new globular clusters. Globulars (such as those around our Galaxy) are traditionally thought of as very old systems, formed alongside the oldest of a galaxy's stars. However, HST images of interacting and merging systems have shown that they can contain numerous bright star clusters which should, when their stellar populations have aged suitably, end up looking just like normal globular clusters. If we believe that many ellipticals have formed via mergers then this can explain why they seem to have so many globular clusters per unit galaxy light compared to their progenitor spirals. (Before the HST observations, this was thought to be a serious *objection* to the formation of ellipticals by the merger of spirals.)

3.10.1   Gravitational interactions

Mergers are best investigated theoretically via computer models of the gravitational forces on individual 'particles'[6] as the galaxies approach one another, the so-called 'N-body models'. Even so we can get some idea of the processes involved via some simple analytic approximations.

To see why galaxies merge, rather than just circle past each other, consider the effect on a star (mass m_*) in one galaxy as a galaxy of mass M_g flies past at velocity v. The star (and galaxy) will gain a small perpendicular velocity due to the attraction of the galaxy (or star) and in this situation we can use the 'impulse approximation'. Measuring the time from the moment of closest approach, at impact parameter b, the

---

[5] Or the even more extreme 'hyperluminous' ones.
[6] Even with current computer technology it is still not possible to simulate all ($\sim 10^{11}$) stars in a large galaxy (though the latest supercomputers used for cosmological simulations can handle a billion particles). A coarser resolution therefore has to be used, each 'particle' representing thousands or millions of solar masses of stars or gas.

distance between the star and the galaxy at time t will be $(b^2 + v^2 \, t^2)^{1/2}$ and the component of force perpendicular to v will be

$$F_\perp = \frac{Gm_* M_g}{(b^2 + v^2 t^2)} \frac{b}{(b^2 + v^2 t^2)^{1/2}} \qquad (3.10.2)$$

This will produce a perpendicular velocity of the galaxy via the acceleration

$$a_\perp = \frac{dv_\perp}{dt} = \frac{Gm_* b}{(b^2 + v^2 t^2)^{3/2}} \qquad (3.10.3)$$

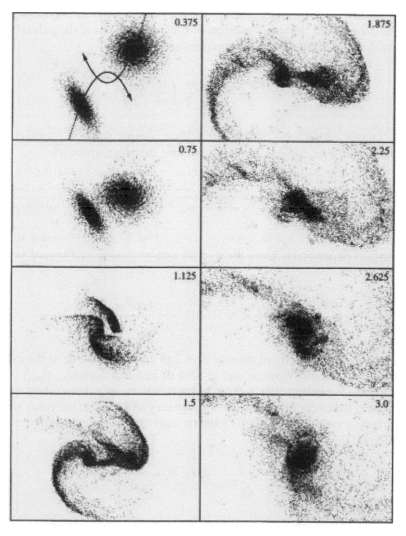

Figure 3.22 Computer simulation of a merger event (Reproduced with permission from Barnes and Hernquist, 1996), showing the distribution of stars at different times.

Integrating over time (from $-\infty$ to $+\infty$) this will result in the galaxy acquiring

$$\Delta\nu_\perp = \frac{2Gm_*}{b\nu} \tag{3.10.4}$$

and the star a corresponding $2GM_g/b\nu$ (by conservation of momentum). Thus we have a gain in perpendicular kinetic energy

$$\Delta K_\perp = \frac{M_g}{2}\left(\frac{2Gm_*}{b\nu}\right)^2 + \frac{m_*}{2}\left(\frac{2GM_g}{b\nu}\right)^2 = \frac{2G^2 m_* M_g(m_* + M_g)}{b^2\nu^2} \tag{3.10.5}$$

with the smaller mass, the star, gaining most of the energy. This energy can only come from the original 'forward' motion of the galaxy. Thus if the galaxy slows by an amount $\Delta\nu$, matching the kinetic energies gives

$$\frac{M_g\nu^2}{2} = \frac{M_g(\nu - \Delta\nu)^2}{2} + \frac{m_*}{2}\left(\frac{M_g}{m_*}\Delta\nu\right)^2 + \Delta K_\perp \tag{3.10.6}$$

where the change in velocity of the star follows simply from the conservation of linear momentum. For $\Delta\nu \ll \nu$ and $m_* \ll M_g$ we therefore arrive at

$$\Delta\nu \simeq \frac{\Delta K_\perp}{M_g\nu} \simeq \frac{2G^2 m_* M_g}{b^2\nu^3} \tag{3.10.7}$$

To obtain the total change in velocity, we need to integrate this over stars at all possible impact parameters. Thus the overall rate of change of forward velocity of the galaxy

$$\frac{d\nu}{dt} = -\int_{b_{min}}^{b_{max}} \frac{2G^2 m_* M_g}{b^2\nu^3} \times n\nu 2\pi b \, db = -\frac{4\pi G^2 m_* M_g n}{\nu^2}\ln\Lambda \tag{3.10.8}$$

where n is the number density of stars, $\nu \, dt \times 2\pi b \, db$ is the volume of the cylindrical shell at impact parameter b swept out in time dt and $\ln\Lambda = \ln(b_{max}/b_{min})$ is what is called a coulomb logarithm. For a close, but not initially interpenetrating, encounter b_{min} could be of the order of the radius of the galaxy and b_{max}, at the other side of the passing galaxy, would be of order 3 galactic radii, so $\ln\Lambda$ would only be of order unity. Roughly, then,

$$\frac{d\nu}{dt} \simeq -\frac{4\pi G^2 M_g\rho_*}{\nu^2} \tag{3.10.9}$$

where $\rho_* = nm_*$ is now the mass density in stars. Notice that we could have done the same calculation for any value of m_* and then summed the contributions at the end, so this final result applies whatever the distribution of star masses within the galaxy.

This deceleration effect is called 'dynamical friction' as it operates in analogous fashion to actual frictional forces (for instance, depending on the galaxy's velocity), and the above is a simplified version of what is called the 'Chandrasekhar formula'. If the deceleration is large enough (which evidently requires small relative velocities and a high mass density of stars), it will eventually slow the passing galaxy to such an extent that it will 'fall in' towards the other galaxy after a number of closer and closer passages. Notice that, somewhat counter-intuitively, larger-mass galaxies are slowed more, so a galaxy will 'swallow' its larger neighbours first (the so-called 'galactic cannibalism'). We can get an idea of the timescale on which this may occur, just from the Chandrasekhar expression, i.e. the time it would take to decelerate from velocity ν to zero at a constant $d\nu/dt$

$$t_{\text{merge}} \sim \frac{\nu}{d\nu/dt} \sim \frac{\nu^3}{4\pi G^2 M_g \rho_*} \tag{3.10.10}$$

For $\nu \sim 200 \, \text{km s}^{-1}$, $M_g \sim 10^{10} M_\odot$ and a density in the central galaxy ρ_* corresponding to $\sim 10^{11} M_\odot$ in a $10 \, \text{kpc}$ radius sphere this gives about 3×10^8 years. In fact, we started our calculation with the galaxies already close together and with a low relative velocity, so the overall time scale for a merger may be somewhat larger. The Large Magellanic Cloud, for example, is expected to merge with the Galaxy in a few $\times 10^9$ years.

Of course this does not tell us the detailed outcome of the process. Perhaps surprisingly, it is found from the computer simulations that even for the merger of two disc galaxies the final product eventually looks just like an ordinary elliptical, with a density distribution mimicking the de Vaucouleurs law[7] with remarkable accuracy.

The overall effects will depend, too, on the ratio of the masses of the two galaxies. A collision with a near-equal-mass galaxy is usually referred to as a major merger, while the 'swallowing' of a smaller companion would be a minor merger or accretion. The dust lanes, polar rings and so forth are then seen as the remains of a spiral galaxy which has 'fallen in', while features like shells or the so-called 'kinematically decoupled cores' (where the central regions rotate independently of the stars further out) probably indicate where the 'relaxation' of the stars in the merged system to an equilibrium state is not yet complete.

3.10.2 Timescales

In fact, on simple dynamical grounds, we might not even expect an isolated elliptical to have reached an equilibrium configuration in the time available since it formed, let alone a merger product. In familiar everyday situations, such equilibration of

---

[7] Nevertheless, Gerard de Vaucouleurs himself remained sceptical, remarking that 'if you collide two automobiles you don't make a new model!'

mixtures of particles is very rapid. If some molecules of oxygen are released into a large container of nitrogen, we know that they will quickly spread out and become completely mixed in; the atmosphere does not consist of separate clumps of nitrogen and oxygen! This occurs because molecules at standard atmospheric temperature and pressure collide with each other about 10^{11} times a second.

By a collision, we really mean that the molecule is more affected by its nearest neighbour than by the averaged-out effects of molecules elsewhere. We can easily adapt this definition for the case of stars in a galaxy. A collision of this sort will completely change the velocities of the stars and cause the same sort of mixing as seen in the gas – providing they occur frequently enough. Such a collision will therefore require potential energy (relative to when the stars were a long way apart) at least of the same order as their kinetic energies, i.e.

$$\frac{Gm_*^2}{r} > \frac{m_* v^2}{2} \tag{3.10.11}$$

where m_* and v are the stars' masses and velocities respectively. For stars around $1M_\odot$ in a giant elliptical, moving at (3-D) velocity $\simeq 500\,\mathrm{km\,s^{-1}}$, this requires

$$r < \frac{2Gm_*}{v^2} \simeq 10^9\,\mathrm{m} \simeq 3 \times 10^{-8}\,\mathrm{pc} \tag{3.10.12}$$

For simplicity consider a star moving straight through the middle of a galaxy for a distance $\sim 2R_e$, say 15 kpc. This will take one 'crossing time'

$$t_{\mathrm{cross}} \simeq 5 \times 10^{20}\,\mathrm{m}/5 \times 10^5 \mathrm{m\,s^{-1}} \simeq 10^{15}\,\mathrm{s} \simeq 3 \times 10^7 \mathrm{years} \tag{3.10.13}$$

During this time it will encounter all the stars in a cylinder of cross-section $\pi r^2 \simeq 3 \times 10^{-15}\,\mathrm{pc^2}$. But we also know that if the column had cross-section $1\,\mathrm{pc^2}$ it would have a column density – and hence number of stars – corresponding to the central surface brightness of a giant elliptical, around $10^5 L_\odot\,\mathrm{pc^{-2}}$, i.e. of order 10^5 stars. Thus in our much thinner column there will be $\sim 3 \times 10^{-10}$ stars, so we would expect one close encounter every 3×10^9 crossing times, or $\sim 10^{17}$ years – very much longer than the age of the universe.

Distant encounters between stars are much weaker, but there will be many more of them. As the change in velocity per encounter will be small, we can use the impulse approximation again and assume a series of short encounters between stars with relative velocity v. With the same notation as before, the component of force perpendicular to v will be

$$F_\perp = \frac{Gm_*^2 b}{(b^2 + v^2 t^2)^{3/2}} \tag{3.10.14}$$

producing a perpendicular velocity

$$v_\perp = \frac{2Gm_*}{bv} \tag{3.10.15}$$

Now in time t the number of stars passing at a distance between b and $b + db$ will be their number density, n, times the volume of the hollow cylinder swept out at distance b, i.e. $2\pi\nu tbdb$. We can thus integrate over b to get the total impulse or change in energy of a star from all other stars passing it in time t,

$$\Delta K = \sum \frac{1}{2}m_*\nu_\perp^2 = \int_{b_{\min}}^{b_{\max}} \frac{1}{2}m_*n\nu t\left(\frac{2Gm_*}{b\nu}\right)^2 2\pi bdb$$

$$= \frac{4\pi G^2 m_*^3 nt}{\nu}\ln\left(\frac{b_{\max}}{b_{\min}}\right) \tag{3.10.16}$$

Relaxation can be taken to occur when $\Delta K \simeq m_*\nu^2/2$, i.e. when the change in energy is equal to the original energy, so that the stars' orbits are well 'mixed'. Thus

$$t_{\text{relax}} = \frac{\nu^3}{8\pi G^2 m_*^2 n \ln \Lambda} \tag{3.10.17}$$

where $\Lambda = b_{\max}/b_{\min}$ as before.

If b covers the range between strong encounters ($\sim 3 \times 10^{-8}$ pc, from above) and the overall size of the galaxy (15 kpc), then $\ln \Lambda = \ln(5 \times 10^{11}) \simeq 27$. The averaged out density n will be of order 10^{11} stars in a volume of 2×10^{12} pc^3, that is about 0.05 pc^{-3}, so

$$t_{\text{relax}} \sim 2 \times 10^{17}\text{yr} \tag{3.10.18}$$

Thus distant encounters cannot significantly drive an elliptical towards equilibrium in a Hubble time, either.

The answer appears to lie in what is called 'violent relaxation', a process first described by Donald Lynden-Bell in 1967. For stars moving in a fixed potential orbital energy is conserved. However, when the galaxy is first forming, by collapse from a larger cloud, or is accumulating mass by mergers or accretion then the gravitational potential clearly is not time-independent. This will change the stars' energies by large, and unpredictable, amounts, mixing them between the various possible orbits, and quite rapidly randomise the velocities. Notice that this is not due to any sort of 'two-body relaxation' between pairs of stars, as in the cases we have just explored, but is a sort of co-operative effect, depending on the changing potential at any given point caused by the motion, and hence change of position, of all the stars in the galaxy (or pair of galaxies).

3.11   Elliptical galaxy masses

From the dynamics of the central regions of ellipticals as discussed in section 3.7, we can deduce that the typical M/L there is around 5 or so. In the case of ellipticals which have detailed observed velocity dispersion profiles (or if we simply assume that they are isothermal), then we can use the same dynamical calculations as we used earlier to find a total mass. We can check this more directly for nearby ellipticals in which velocities of individual 'stars' (generally, in fact, planetary nebulae) can be measured. This provides an overall estimate of M/L of order 10 or more. On the other hand, for globular clusters, which look like similar collections of old stars, the velocity dispersions indicate that M/L is more like 1–2. In addition, if we calculate the light output for a certain total mass in stars, whose individual masses are distributed according to a realistic IMF – for example, by using one of the population synthesis models we mentioned in section 3.3 – we find that an old galaxy should have an M/L of around 2–3. Thus ellipticals seem to have rather more mass than we might expect.

If the galaxy has a number of satellites, either globular clusters, if the galaxy is near enough for us to distinguish them, or dwarf galaxies, then we may be able to use the dynamics of these objects to determine the mass out to very large scales. As a minimum, we will require the galaxy to be sufficiently massive that the satellites are moving slower than the local escape velocity at the distance R_{sat} from the centre of the galaxy

$$v_{esc}^2 \simeq \frac{2GM(R_{sat})}{R_{sat}} \tag{3.11.1}$$

Calculating the required masses for satellites at radial distances around 100 kpc, the inferred M/L can be of order 50.

As another alternative, we can use the X-ray-emitting gas that we met in section 3.7 to delineate the gravitational potential. We can use the standard equation of hydrostatic equilibrium

$$\frac{dP}{dr} = -\rho(r)\frac{GM(r)}{r^2} \tag{3.11.2}$$

where P is the pressure and ρ the density of the gas at radius r, and as usual $M(r)$ is the mass inside r. For an isothermal gas with an average particle mass μm_p, where m_p is the proton mass (so that for completely ionised hydrogen with a typical admixture of helium $\mu \simeq 0.6$),

$$P = nk_BT = \frac{\rho}{\mu m_p}k_BT \tag{3.11.3}$$

Now if the particle density $n \propto r^{-\alpha}$, then the emission per unit volume will be $\epsilon \propto n_p^2 \propto r^{-2\alpha}$. Projecting this will give an X-ray surface brightness, $I_X \propto R^{1-2\alpha}$, so we can work back from the observed X-ray profile to $n_p(r)$ and hence $\rho(r)$. We can also relax the assumption of isothermality if we have a spatially resolved X-ray spectrum to give us $T(r)$. Finally we can invert the hydrostatic equation to give

$$M(r) = \frac{k_B}{G\mu m_p}\frac{r^2}{\rho}\left(-\frac{d(\rho T)}{dr}\right) \tag{3.11.4}$$

For the isothermal case and an observed power law surface brightness profile, this reduces to

$$M(r) = \frac{\alpha k_B T}{G \mu m_p} r \tag{3.11.5}$$

Large central cluster ellipticals, for instance, can contain $\sim 10^{11} M_\odot$ of X-ray gas (comparable to their mass in stars), but the total dynamical mass, out to the limits of the X-ray observations, is typically an order of magnitude larger.

The general increase of M/L with radius, and the fact that the calculated M/L is far higher than expected for any population of stars, even inside the main luminous body of the galaxy, is one manifestation of the famous 'missing mass problem', that is the inferred presence of gravitating mass which cannot be accounted for by visible (stellar plus gaseous) matter. The problem was first raised in the late 1930s, independently by Smith and Zwicky. They found that the dynamical masses of galaxy clusters (which are dominated by giant ellipticals, remember), estimated in exactly the same way as we have done for individual galaxies, were far larger than could be accounted for by the visible galaxies if they had M/L of a few. (Individual galaxy masses had not been measured at that point.) This 'dark matter' responsible for the high M/L is now known not to be made of ordinary atoms, etc. of any sort, but to come in some exotic (but still unspecified) form of 'non-baryonic' matter. We will consider the dark matter in more detail in Chapters 4 and 7.

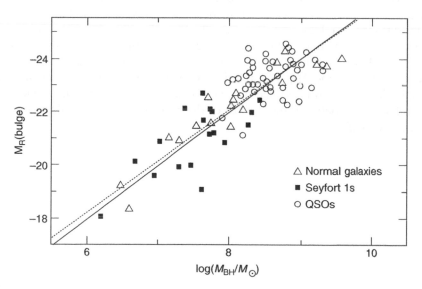

Figure 3.23 The relationship between central black hole mass and the luminosity of the spheroidal component of its host galaxy (Reproduced with permission from McLure and Dunlop, 2002).

3.12 Massive black holes

If we move right into the very centres of galaxies, then we can use similar methods to study the mass in the nuclear region. Observations of the velocity dispersion at high resolution, particularly with HST, have revealed that σ increases as we approach the centre. Keeping these rapidly moving stars close to the centre clearly requires a deep potential well and therefore a large central mass. The Local Group (dwarf) elliptical M32 was quite early on found to require a central mass of around $10^6 M_\odot$ on parsec scales. In genuine giant ellipticals like M87, the velocity dispersion implies that there must be a few times $10^9 M_\odot$ inside the central 10 pc, corresponding to a density above $10^6 M_\odot \, \mathrm{pc}^{-3}$. M87 also contains a very small central gas disc, again observable with HST, whose rotation velocity corroborates the large central mass. This is hugely in excess of what can be accounted for by visible stars or gas and is so concentrated that it is virtually certain that the cause must be a supermassive black hole (SMBH).

With increasing numbers of observed systems, it has been found that the inferred black hole mass, M_{BH}, correlates well with the luminosity of the elliptical host (or, including other galaxy types, with the bulge luminosity). An even tighter relationship is found between black hole mass and overall galaxy velocity dispersion, viz.

$$M_{\mathrm{BH}} \simeq 1.4 \times 10^8 \left(\frac{\sigma_0}{200 \, \mathrm{km \, s^{-1}}} \right)^\beta M_\odot \qquad (3.12.1)$$

with β around 3.5 (but perhaps steepening at low mass).

Recall that some ellipticals are strong radio sources and their radio emission is almost certainly powered by accretion onto a central black hole. However, it appears that there is little correlation between the black hole mass and whether an elliptical is radio loud (in the usual sense of the term; all ellipticals are presumably radio emitters at some level), except that radio emission seems to require a black hole of mass at least $10^9 M_\odot$. For example, Cygnus A is a very powerful radio source but its host galaxy sits on the same M_{BH}–L_B relation as non-active elliptical. This apparently universal relationship, often called the 'Magorrian relation', is taken as evidence for a close relationship between the growth of the central black hole and the formation of the galaxy itself.

4 Spiral galaxies

4.1 Shapes and sizes

The other major variety of large galaxy is, of course, the spiral. These are usually found in relatively low density environments compared to ellipticals, typically in structures akin to the Local Group. As we saw in Chapter 2, spirals come in a wide variety of forms. All are characterised by (relatively) thin, (more or less) flat discs, and most have a central bulge which is in many ways like an elliptical galaxy. The importance of the bulge, usually quantified by the bulge-to-disc ratio B/D (in terms of total luminosity), spans a large range. The earliest type spirals, and the S0s, have the largest B/D, typically around unity, with B/D then decreasing to around 0.3 for Sb galaxies similar to our own and M31, and to less than 0.1 for later type spirals like M33. Sd and Sm galaxies, like the LMC, have no real bulge at all.

Using the radial intensity profiles from section 2.4, we can see that a general spiral galaxy should have an intensity profile

$$I(R) = I_{0D} \exp(-R/a_D) + I_{0B} \exp\left(-(R/a_B)^{1/4}\right) \qquad (4.1.1)$$

where the extra subscripts 'B' and 'D' refer to the bulge and disc components. Unfortunately we cannot write the (logarithmic) surface brightness in magnitudes

The Structure and Evolution of Galaxies   Steven Phillipps
© 2005 John Wiley & Sons, Ltd

in any simple way as we could for the disc and bulge separately. Using the equations derived in section 2.4 we can, though, write the bulge-to-disc ratio as

$$\frac{B}{D} = 3.5 \frac{R_{eB}^2 I_{eB}}{a_D^2 I_{0D}} \qquad (4.1.2)$$

The bulges themselves have physical sizes, as measured by their effective radii, of anything from 100 pc up to several kpc, similar to the size of moderately luminous ellipticals. Their forms vary; they can be fairly round spheroids, or quite flattened or even prolate tri-axial ellipsoids. When seen edge-on, most bulges have elliptical iso-photes; but some are 'peanut shaped', with the isophotes showing a dip at the minor axis.

Bulges share many dynamical properties with the lower luminosity elliptical galaxies. They have substantial velocity dispersions ($\simeq 110\,\mathrm{km\,s^{-1}}$ for the Galaxy), but also significant rotation, with generally $V/\sigma \sim 1$. The shapes of bulges are consistent with their being rotationally flattened.

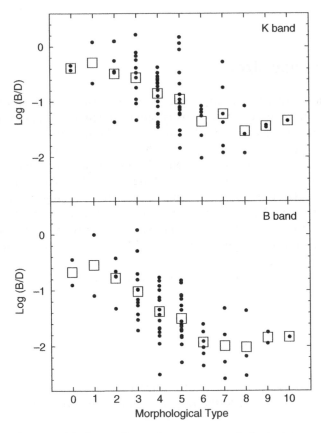

Figure 4.1 Distribution of bulge-to-disc ratio versus morphological type, in the infra-red (top) and blue (bottom) (from Graham, 2001).

The other component in the inner region of some galaxies is a bar. Anything up to half of all disc galaxies are barred. S0 galaxies are apparently just as likely to be barred as 'real' spirals, so clearly the presence of gas is not necessary to their existence, unlike the spiral structure which we discuss later in the chapter. In early type barred spirals, the bar appears to co-exist with the bulge. The bars themselves may be quite elongated, with an axis ratio up to 5 and can contain as much as 30% of the total galaxy light. They are also quite flat, like the disc component. Bars rotate more or less as solid bodies, with a 'pattern speed' Ω_p, in the sense that the entire bar will point in different directions at different times. The individual stars move on long thin orbits (as seen from the frame rotating at angular speed Ω_p), so as to stay within the bar.

4.1.1 Spiral arms

While the bulge decreases in prominence along the Hubble sequence, the structure of the disc varies, too. The arms of Sa galaxies are very tightly wound, but in later types the arms open out. The shape of the spiral pattern can be represented by a logarithmic spiral. In polar coordinates this has the form

$$\ln(R/R_0) = k\theta \qquad (4.1.3)$$

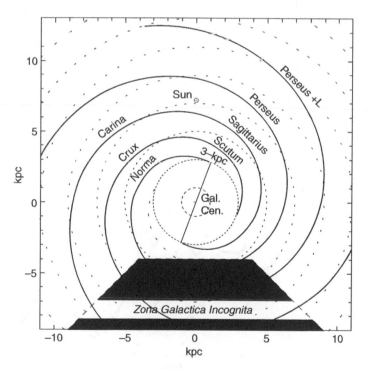

Figure 4.2   The pitch angle of a spiral arm is the angle between the tangents to the arm (solid line) and the local azimuthal direction (dotted circles) (Reproduced with permission from Vallée, 2002).

where R_0 is the 'radius' of the spiral as it crosses the major axis, at $\theta = 0$. Of course, a pure mathematical spiral is infinitely long and has infinitely many crossings of $\theta = 0$. In practice, for a spiral galaxy, the arms start from the edge of the bulge or the ends of the bar and complete no more than a couple of turns around the galaxy.

The 'pitch angle', i, measures the angle between the direction of the arm (or strictly the tangent to it) and the tangent to a circle at the same R, so we have

$$\tan i = \frac{\mathrm{d}R}{R\mathrm{d}\theta} \qquad (4.1.4)$$

Hence for a logarithmic spiral

$$i = \tan^{-1} k \qquad (4.1.5)$$

For Sa galaxies i is around $5°$, for mid-type spirals like ours it is around $10°$ to $12°$ and for late types $i \sim 20°$.

We usually think primarily in terms of two-armed spirals – so that at any radius R there are two brightness peaks $180°$ apart – but in fact spirals can have multiple arms. If the two, or sometimes four, arms are continuous over large azimuth ranges, say $270°$ or more, the spiral is much more impressive and is referred to as a 'grand design' spiral. The opposite, where the arms come in multiple short fragments or have numerous branching sub-arms, are called 'flocculent' spirals. The latest Sm types have just rudimentary structure, only approximating to arms.

The arms are superimposed on the exponential disc of starlight which, as we have seen, may or may not dominate over the central bulge. Despite appearances, the arms themselves are a relatively minor component in terms of the total number of stars. A cross-section through a spiral galaxy, say in the infra-red K-band which is

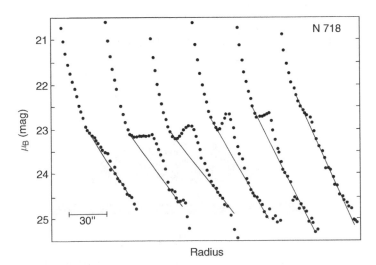

Figure 4.3 Radial intensity profiles (at various position angles), showing cross sections through the main spiral arm of NGC 718 (Reproduced with permission from Kennicutt and Edgar, 1986).

insensitive to the presence of a few very bright blue stars, shows only a fairly small intensity enhancement as we cross the arms.

4.1.2   Surface brightnesses

The disc components of bright spirals have scale lengths of typically a few kpc, for instance M31 has $a_D \simeq 4.5$ kpc. Recall that for exponentials, $r_e = 1.69a$, so the half light radii of discs can range up to 10 kpc or so. The overall observable extent can be many times larger than this. For instance we can characterise it by the extent at which the surface brightness drops to 25 Bμ (the de Vaucouleurs radius, as used in de Vaucouleurs' *Reference Catalog of Bright Galaxies*[1]). This is typically about $3a_D$. Going to even fainter limits, the Holmberg radius is that at roughly the 26.5 Bμ isophote (essentially the faintest detectable with photographic observations) and is generally around 4–5a_D, so perhaps 25 kpc for bright spirals. Some discs appear to end quite suddenly at this sort of radius, rather than just fade away into the background. (Recall that 26.5 Bμ is only a couple of per cent of the sky background brightness, so the contrast of the galaxy image against the sky is very poor at these levels.)

Although the central regions of the disc are hidden by the brighter bulge in many cases, we can nevertheless extrapolate the radial profile that we see outside the bulge

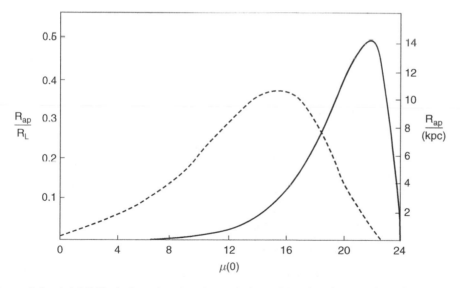

Figure 4.4   A 'visibility' plot, showing the variation of isophotal size of a galaxy of fixed luminosity as the central surface brightness is varied. Calculations for both disc (solid line) and bulge/elliptical (dashed) profiles are shown (Reproduced with permission from Disney and Phillipps, 1983).

---

[1] Actually there are three catalogues (with various co-authors) known for short as the RC1, RC2 and RC3. Before the advent of data centres, the World Wide Web, etc., these were the prime repositories for quantitative information on galaxies.

(i.e. where the disc dominates) to obtain an estimate of the disc brightness at the centre, the extrapolated central surface brightness, μ_{0D}. Many well-known large bright spirals appear to have closely similar extrapolated values of $\mu_{0D} \simeq 21.65\,\text{B}\mu$ (corresponding to just over $100 L_\odot\,\text{pc}^{-2}$). This is known as 'Freeman's law'. However, this may be another example of a selection effect, just as we met with the prevalence of very luminous galaxies (or stars) amongst the brightest ones on the sky.

Consider a set of disc galaxies all with the same total luminosity but different central surface brightnesses. Now for fixed L_T we must have smaller scale sizes for higher surface brightnesses. Thus if we have a galaxy of very high surface brightness, it will be physically very small and its isophotal radius, say at the $25\,\text{B}\mu$ isophote, will also be small. If on the other hand, the surface brightness is very low, then only the very central region will be above the limiting isophote and again the isophotal radius will be small. There will therefore be a preference for galaxies of some intermediate surface brightness in the sense that this lends the galaxy its maximum possible apparent size.

In fact we can quantify this quite easily, as first pointed out by Mike Disney in 1976. The isophotal radius R_{lim} at some limiting or threshold intensity I_{lim} is given by

$$I_{\text{lim}} = I_0 \exp(-R_{\text{lim}}/a) \tag{4.1.6}$$

(For convenience we have suppressed the subscript 'D'). Inverting this,

$$R_{\text{lim}} = a \ln(I_0/I_{\text{lim}}) = 0.92a(\mu_{\text{lim}} - \mu_0) \tag{4.1.7}$$

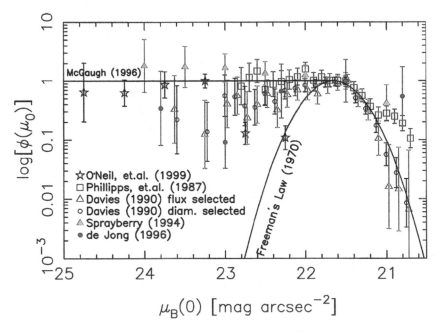

Figure 4.5 A compilation of surface brightness distributions (from O'Neil and Bothun, 2000) showing the large uncertainties and discrepancies between different authors at the faint μ_0 end.

But also $L_T = 2\pi a^2 I_0$, so we have, for fixed L_T,

$$R_{\lim} \propto I_0^{-\frac{1}{2}} \ln(I_0/I_{\lim}) \propto (\mu_{\lim} - \mu_0) 10^{0.2(\mu_0 - \mu_{\lim})} \qquad (4.1.8)$$

As expected, this is a peaked function of μ_0, and with the peak occuring around 2.2 magnitudes brighter than the limiting isophote. With older photographic imaging ($\mu_{\lim} \simeq 24\,B\mu$) this peak therefore closely coincides with the value picked out by Freeman's law.

Besides possibly influencing the choice of large spiral galaxies on which to carry out detailed photometry, this bias can also affect the inclusion of different types of galaxy in overall samples, both photometric and spectroscopic. If we include as galaxies only those objects whose images are above some minimum angular diameter on our detector, then clearly galaxies (of a given L) which present large physical isophotal sizes, as above, will be included out to much larger distances and hence in relatively large numbers. Galaxies with very high or very low surface brightnesses will exceed the angular diameter limit only if they are quite local. Indeed, *very* low surface brightness galaxies may not rise above the survey's surface brightness threshold at all. It is easy to see that a related effect works on isophotal magnitudes, too. Again, low surface brightness galaxies will have a smaller fraction (possibly none) of their total light inside the limiting isophote, so are less likely to be counted in a given survey. This is particularly pertinent for the large redshift surveys, where it is clearly only the detectability down to some threshold which will generate inclusion.

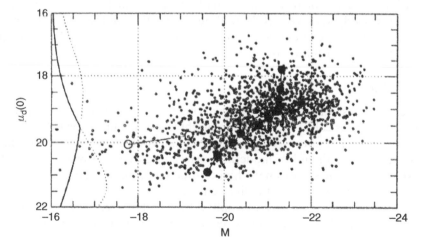

Figure 4.6 The BBD for a sample of Sc galaxies, observed in the I band (Reproduced with permission from Giovanelli *et al.*, 1995).

Recent work, taking into account this type of bias (sometimes referred to as the 'visibility' of different types of galaxy), indicates that the decline in numbers on the bright side of Freeman's peak is genuine (though some high surface brightness discs do exist). However, on the other side, there is a whole population of LSBGs which have been uncovered in the last decade or so by deeper photometric surveys, like the

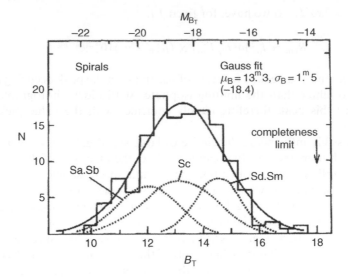

Figure 4.7　The LFs for various spiral types in the Virgo Cluster (Reproduced with permission from Sandage, Binggeli and Tammann, 1985).

photographic Second Palomar Observatory Sky Survey (POSS-II) or the many Charge Coupled Device (CCD) based surveys. Indeed, some work suggests that the distribution of μ_0 for discs may actually be quite flat from $\simeq 21.5\,\mathrm{B}\mu$ down to the faintest properly surveyed levels at about $25\,\mathrm{B}\mu$, though this remains controversial; according to the recent large redshift surveys, bright spirals, at least, do appear to have a peaked distribution of μ_0.

4.1.3　Numbers

We can extend these discussions by considering the joint distribution of μ_0 and a or μ_0 and L (sometimes called the 'bivariate brightness distribution' or BBD, as it is the distribution in total brightness and surface brightness). It turns out that the majority of discs are quite small and of moderate surface brightness. However, as the large, higher surface brightness discs have higher total luminosities, the large galaxies are the most important in terms of the light emitted.

If we return to the ordinary LF and split it by morphological types, as discussed in section 3.1, we find that spirals (and irregulars) have Schechter function parameters

$$\alpha \simeq -1.2$$

$$M_* \simeq -20.6$$

$$\phi_* \simeq 0.004$$

Splitting this further, the late types have fainter M_* but steeper α than the earlier types. Thus a typical Sd, say, is significantly less luminous than an average Sa. The

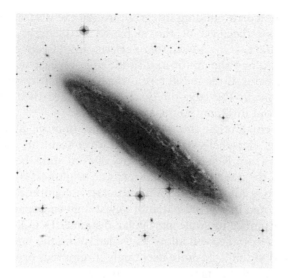

Figure 4.8   The edge-on galaxy NGC 253. (UKST image, © AAO.)

fractions of galaxies in each spiral type are relatively poorly defined; but roughly speaking, amongst brighter spirals there are similar numbers of Sa and Sb galaxies combined as there are of Sc types. At the fainter end Sd and Sm galaxies start to dominate over the earlier types.

4.2   Vertical structure

Of course, the discs of spirals are not infinitely thin. If we observe edge-on examples, then we see a flattening of up to a factor 10. We therefore expect that physical disc thicknesses should be of order 10% of their radii. More quantitatively, the shape we see must depend on the vertical structure of the disc. We already know that the radial (projected) surface density of face-on discs is well fitted by an exponential law and if we look at the minor axis profile of an edge-on galaxy we find that we can use the same form in the vertical (z-) direction. This suggests that, in cylindrical polars, we can write the (3-D) density structure in the (cylindrically symmetric) form

$$\rho(R, z) = \rho_0 \exp(-R/a) \exp(-|z|/h) \tag{4.2.1}$$

with typically the scale height $h \simeq 1/10$ of the scale length a, though there do exist some so-called 'super-thin' galaxies with much smaller scale heights.

Notice that because the distribution is separable in R and z (unlike the case for ellipsoidal density distributions), projection onto the sky (nearly always) leaves the form unchanged. For instance, it is easy to check that if we place such a galaxy face-on to the observer then the projection is the density at some R integrated over all z, which just introduces a multiplier $2h$; for example, the central surface brightness I_0 corresponds to $2h\rho_0$. In fact, it is straightforward to see that at intermediate inclination angles the corresponding integral through the disc is just increased by a further factor $1/\cos i$ (because of the increased pathlength), again leaving the exponential form intact. (Of course, this neglects any effects of dust in the disc, section 4.9.) Finally, along the minor-axis in the edge-on case we get $\rho_0 \exp(-|z|/h)$ multiplied by the integral of the radial term, which is $2a$. The exception to the rule is the major-axis profile in the edge-on case, since we are then integrating across different ranges of R at different points along the axis.

In our Galaxy we can analyse the vertical structure in more detail via star counts, an updated version of Herschel's star gauging. If we concentrate on a particular type of star, say main-sequence K stars (the majority of faint stars, below $m_V = 14$, are K or M 'dwarfs'), all of them will have similar intrinsic brightnesses, around half a solar luminosity. But if they all have the same absolute magnitude, $M_V \simeq 6$, then the number of stars counted in a given patch of sky at a given apparent magnitude will correspond to the number at some particular distance. Detailed spectroscopy, or accurate $B-V$ or $V-I$ colours, will allow us to determine the luminosity, and hence distance, of each individual star more accurately if required, or we can go the other way and merely count all the stars at a given apparent magnitude. In the latter case, what we get will be a convolution of the density variation along the line of sight, $\rho(D)$, with the stellar luminosity function $\phi(M)$, since stars at a given m can come from any suitable combination of M and D, i.e.

$$n(m) = \int \phi(M = m - 5\log(D/10) + A(\ell, b, D))\, \rho(D)\, \Omega D^2 \mathrm{d}D \qquad (4.2.2)$$

where A here is the absorption out to distance D in direction (ℓ, b) and Ω is the solid angle surveyed. (Strictly, this will only be precisely true if the LF does not change with distance along the line of sight, as it might if we are observing a mixture of populations.)

4.2.1 Thin and thick discs

Observations towards the Galactic Poles show that for K dwarfs the vertical distribution near the Galactic Plane is indeed close to exponential, with a scale height around 350 pc. For A stars, though, the scale height is only about 200 pc and the youngest, very luminous, stars lie in an even thinner layer. Looking in more detail at the K dwarfs, we see that there is a 'tail' to higher z distances. This is the signature of the 'thick disc', first identified by Gerry Gilmore and Neil Reid in 1983, with a scale height around 1350 pc. Further out still, we see another component starting to dominate. This is the halo, a diffuse spheroidal extension of the bulge, with stellar density fall-off $\sim r^{-3}$. In the volume within, say, 10 pc of the Sun, called the Solar Neighbourhood, around 98% of stars belong to the thin disc, $\simeq 2\%$ to the thick disc and possibly $\sim 0.1\%$ are halo stars.

The older stars have a broader distribution, even in the thin disc itself, due to the scattering effect, over time, of near collisions with other stars or, more importantly, with much more massive Molecular Clouds. As would be expected, the stars which have had time to be scattered the most have higher vertical velocities and can travel further out of the Plane. The Sun, for instance, has a vertical velocity ('upwards', i.e. towards the North Galactic Pole) of about 7 km s^{-1}. This is actually quite a small

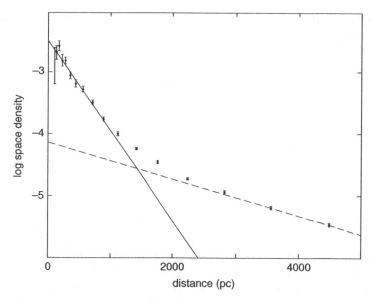

Figure 4.9 The vertical structure of the Galaxy, determined from star counts towards the South Galactic Pole (Reproduced with permission from Gilmore and Reid, 1983).

value, with many thin disc stars having z-velocities around $20\,\mathrm{km\,s^{-1}}$. Thick disc stars have $v_z \simeq 40\,\mathrm{km\,s^{-1}}$.

4.2.2 Surface densities

Comparing the vertical velocities to the vertical extent, we can estimate the surface mass density; stars will move out to greater distances for smaller restoring forces. Specifically, if we treat some class of stars as tracer particles, with number density n, then the collisionless Boltzmann equation of statistical mechanics tells us that the densities and velocities are related by

$$\frac{\mathrm{d}}{\mathrm{d}z}\left(n(z)\sigma_z^2\right) + \frac{\partial\Phi}{\partial z}n(z) = 0 \tag{4.2.3}$$

where Φ is the gravitational potential, and therefore $\partial\Phi/\partial z$ represents the vertical force, often called K_z.

But we also have Poisson's equation relating the mass density ρ to the potential. In cylindrical polars this is

$$4\pi G\rho(R,z) = \nabla^2\Phi(R,z) = \frac{\partial^2\Phi}{\partial z^2} + \frac{1}{R}\frac{\partial}{\partial R}\left(R\frac{\partial\Phi}{\partial R}\right) \tag{4.2.4}$$

Also, matching the radial force to the centripetal acceleration, we can write

$$R\frac{\partial\Phi}{\partial R} = V^2(R) \tag{4.2.5}$$

where V is the circular velocity (i.e. the velocity in a circular orbit at radius R). As this is observed to be very nearly constant in the outer parts of most spiral discs (section 4.14), the R derivative term in equation (4.2.4) is generally very small. Thus

$$4\pi G \rho(R, z) \simeq \frac{\mathrm{d}}{\mathrm{d}z}\left[-\frac{1}{n(z)}\frac{\mathrm{d}}{\mathrm{d}z}\left(n(z)\sigma_z^2 \right) \right] \tag{4.2.6}$$

Assuming the disc is symmetrical about the midplane, we can integrate this over all z to obtain the surface density Σ from

$$2\pi G \Sigma(R) = 2\pi G \int_{-\infty}^{\infty} \rho(R, z)\mathrm{d}z \simeq -\frac{1}{n(z)}\frac{\mathrm{d}}{\mathrm{d}z}\left(n(z)\sigma_z^2 \right) \tag{4.2.7}$$

Jan Oort first used this method for determining the surface density near the Sun in 1932, using F dwarf and K giant stars as tracers. More recent work has used the more numerous K dwarfs. Allowing for a gradual increase in σ_z^2 with height, from $\simeq 20\,\mathrm{km\,s}^{-1}$ to $30\,\mathrm{km\,s}^{-1}$, the observed $n(z)$ implies $\Sigma \simeq 60 \pm 10 M_\odot\,\mathrm{pc}^{-2}$ (integrated out to $\simeq 1\,\mathrm{kpc}$; above that we would be including a contribution from the halo). This is, as Oort had already discovered, greater than the density in visible stars. However, allowing for faint white dwarfs and low mass red and brown dwarfs, the stellar contribution may reach $40 M_\odot\,\mathrm{pc}^{-2}$ and there is a contribution of some $12 M_\odot\,\mathrm{pc}^{-2}$ from gas in the interstellar medium, so around $50 M_\odot\,\mathrm{pc}^{-2}$ may be directly accounted for.[2] The amount of dark matter in the disc is therefore probably quite small.

Note that if all the mass was concentrated in a uniform thin sheet of surface density Σ, the potential at height z would be $2\pi G \Sigma z$, so for fixed σ_z^2 we would have

$$\sigma_z^2 \frac{\mathrm{d}n}{\mathrm{d}z} = -n\frac{\partial \Phi}{\partial z} \tag{4.2.8}$$

implying

$$n = n_0 \exp\left(-2\pi G \Sigma z / \sigma_z^2 \right) \tag{4.2.9}$$

i.e. an exponential distribution with scale height $h = \sigma_z^2 / 2\pi G \Sigma$. For the above densities and velocities, we obtain $h \sim 200\,\mathrm{pc}$, consistent with observation.

4.3   Rotation

The other main dynamical feature of discs, of course, is that they rotate. It is relatively straightforward to measure the disc rotation in external disc galaxies, especially if they are nearly edge-on, as we merely require to determine the Doppler shift of lines in the galaxy's spectrum at different points along the major axis.

---

[2] This translates to about 100 grams per square metre, the same surface density as a sheet of paper!

Generally speaking, the rotation speed rises fairly quickly (from zero at the centre, obviously) in the inner galaxy before levelling off. Early type spirals with the largest bulges have the most rapid rise. Recall from simple motion in a circle of radius R (like planets around the Sun) that matching the required centripetal acceleration by the gravitational force implies

$$V^2 = \frac{GM}{R} \tag{4.3.1}$$

where, for an extended mass, the relevant quantity is $M(R)$, the mass inside radius R. Strictly, this only applies to spherically symmetric mass distributions, but it is still a good approximation even for flattened distributions. If the mass is centrally concentrated, like the light appears to be, we might expect to be able to approximate this at large R by just

$$V^2 = \frac{GM_T}{R} \tag{4.3.2}$$

where M_T is the total mass. Thus we should get 'Keplerian' orbits with $V \propto R^{-1/2}$. There is no real sign of this decline in velocity in the optical spectra, but until quite recently observations were limited to fairly small radii, anyway, because of the fall in surface brightness with radius.[3] Extending the 'rotation curves' to greater galacto-centric distances therefore depended on finding another tracer, as we shall see in section 4.14.

In our Galaxy, we might hope to be able to measure the rotation curve $V(R)$ with much better resolution. However, our position inside the rotating disc makes this rather harder than one might initially think. Nevertheless, consider a star at some particular Galactic longitude ℓ, a distance D from the Sun and at Galacto-centric radius R. If we measure the radial velocity of the star, we can infer the rotational velocity $V(R)$. Ignoring non-circular motions for the moment, let us suppose that the region of the Galaxy near the Sun is orbiting with speed V_0. (The Sun itself moves relative to this average local motion – the Local Standard of Rest (LSR) – at a few km s^{-1}, but we can allow for this deviation).

From Figure 4.10 we can see that the star will have a radial velocity

$$\nu_r = V \cos \alpha - V_0 \sin \ell \tag{4.3.3}$$

Also, from the sine rule in the triangle Sun–star–Galactic Centre,

$$\frac{\sin \ell}{R} = \frac{\sin(\pi/2 + \alpha)}{R_0} \tag{4.3.4}$$

---

[3] Though as far back as 1939, Horace Babcock deduced that the outer regions should contain significant mass 'on the basis of the unexpectedly large circular velocities in these parts'.

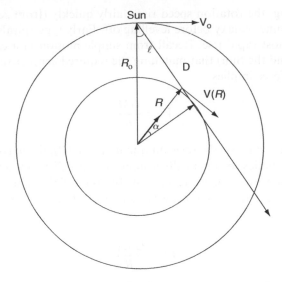

Figure 4.10　Illustration of the geometry of Galactic rotation as seen from the Sun.

so

$$\nu_r = R_0 \sin \ell \left(\frac{V}{R} - \frac{V_0}{R_0} \right) \qquad (4.3.5)$$

If the disc rotated as a solid body (i.e. $V \propto R$) then there would be no relative velocity ν_r. However, as we have seen, outside the central regions, discs show differential rotation, stars further out taking longer to orbit. In this case, nearby stars in the 'first quadrant' ($0 \leq \ell \leq 90°$) appear to move away from us (positive ν_r), whereas distant stars in the same direction are approaching us (negative ν_r). In the outer Galaxy, ν_r is always negative in the second quadrant (90° to 180°) and always positive in the third. Finally, in the quadrant $270° \leq \ell \leq 360°$, it is the distant stars which are moving away and the nearby ones which are approaching.

4.3.1　Oort's constants

In principle, if we could observe stars of known ℓ and D throughout the disc, we could determine the entire $V(R)$ directly. However, the disc is full of obscuring dust in the ISM, especially towards the Galactic Centre, so this is not really a practicable proposition. The rotation curve can therefore only be determined accurately, from stars, for regions quite close to the Sun.

In this case, i.e. $D \ll R_0$, we have approximately

$$R = R_0 - D \cos \ell = R_0 + \Delta R \qquad (4.3.6)$$

The radial velocity then becomes

$$\nu_r = R_0 \sin \ell \frac{\mathrm{d}}{\mathrm{d}R} \left(\frac{V}{R}\right) \Delta R$$

$$= -\frac{R_0 D}{2} \sin 2\ell \frac{\mathrm{d}}{\mathrm{d}R} \left(\frac{V}{R}\right)_{R_0}$$

$$= -\frac{R_0 D}{2} \sin 2\ell \left[\frac{1}{R}\frac{\mathrm{d}V}{\mathrm{d}R} - \frac{V}{R^2}\right]_{R_0}$$

$$= \frac{D}{2} \sin 2\ell \left[\frac{V_0}{R_0} - \left(\frac{\mathrm{d}V}{\mathrm{d}R}\right)_{R_0}\right] \tag{4.3.7}$$

If we now define 'Oort's constant A' via

$$A = -\frac{R_0}{2}\frac{\mathrm{d}}{\mathrm{d}R}\left(\frac{V}{R}\right)_{R_0} = \frac{1}{2}\left(\frac{V}{R} - \frac{\mathrm{d}V}{\mathrm{d}R}\right)_{R_0} \tag{4.3.8}$$

then we have

$$\nu_r - AD \sin 2\ell \tag{4.3.9}$$

A is a measure of the gradient of the rotation curve near the Sun and observationally has a value of about $14 \, \mathrm{km \, s^{-1} \, kpc^{-1}}$.

The tangential velocity of the star, measured from its proper motion (recall equation 1.5.1), is similarly

$$\nu_t = \frac{R_0 D}{2} \cos 2\ell \frac{\mathrm{d}}{\mathrm{d}R}\left(\frac{V}{R}\right)\Delta R - \frac{D}{2R_0}\frac{\mathrm{d}}{\mathrm{d}R}(RV)\Delta R$$

$$= D(A \cos 2\ell + B) \tag{4.3.10}$$

where we have defined Oort's constant B by

$$B = -\frac{1}{2}\left[\frac{1}{R}\frac{\mathrm{d}}{\mathrm{d}R}(RV)\right]_{R_0} = -\frac{1}{2}\left(\frac{\mathrm{d}V}{\mathrm{d}R} + \frac{V}{R}\right)_{R_0} \tag{4.3.11}$$

Note that B is negative and takes a value around $-12 \, \mathrm{km \, s^{-1} \, kpc^{-1}}$. Physically it is a measure of the angular momentum gradient.

Notice that

$$A + B = -\left(\frac{\mathrm{d}V}{\mathrm{d}R}\right)_{R_0} \tag{4.3.12}$$

and

$$A - B = \frac{V_0}{R_0} \tag{4.3.13}$$

The measured values of Oort's constants therefore imply that V is decreasing slightly with radius near the Sun; a 'flat' rotation curve with constant V would have $B = -A$ exactly. Also, assuming that the Sun is about 8.5 kpc from the Galactic Centre, we must have $V_0 \simeq 220 \, \text{km s}^{-1}$.

4.3.2   Epicyclic motions

The motions of the stars are not purely circular, for instance we have noted that the Sun moves relative to the LSR at a few km s^{-1} in each direction. For an axisymmetric galaxy we can write the gravitational potential as $\Phi(R,z)$ and note that the angular momentum about the z-axis will be conserved for any particular star,

$$L_z = R^2 \frac{\mathrm{d}\theta}{\mathrm{d}t} = \text{constant} \tag{4.3.14}$$

The equation of motion in the radial direction depends on the difference between the centrifugal and gravitational terms,

$$\frac{\mathrm{d}^2 R}{\mathrm{d}t^2} = R\left(\frac{\mathrm{d}\theta}{\mathrm{d}t}\right)^2 - \frac{\partial \Phi}{\partial R} = -\frac{\partial}{\partial R}\left(\Phi + \frac{L_z^2}{2R^2}\right) \tag{4.3.15}$$

Multiplying each side by $\mathrm{d}R/\mathrm{d}t$ and integrating shows that for a given star

$$\frac{1}{2}\left(\frac{\mathrm{d}R}{\mathrm{d}t}\right)^2 + \Phi + \frac{L_z^2}{2R^2} = \text{constant} \tag{4.3.16}$$

Similarly, the vertical motion is governed by

$$\frac{\mathrm{d}^2 z}{\mathrm{d}t^2} = -\frac{\partial \Phi}{\partial z} \tag{4.3.17}$$

For nearly circular orbits at R_* in a thin disc, we can use a Taylor series expansion in R and z about the point $(R_*, 0)$,

$$\Phi(R,z) \simeq \Phi(R_*, 0) + \frac{\partial \Phi}{\partial R}(R - R_*) + \frac{\partial \Phi}{\partial z} z \tag{4.3.18}$$

We then have

$$\frac{\mathrm{d}^2 z}{\mathrm{d}t^2} \simeq -z\left[\frac{\partial^2 \Phi}{\partial z^2}(R_*, z)\right]_{z=0} \tag{4.3.19}$$

So if we write

$$\omega^2 = \left[\frac{\partial^2 \Phi}{\partial z^2}\right]_{z=0} \qquad (4.3.20)$$

we obtain

$$\frac{d^2 z}{dt^2} = -\omega^2(R_*)z \qquad (4.3.21)$$

which is just the usual equation for simple harmonic motion with angular frequency ω. Thus

$$z = z_1 \sin \omega t \qquad (4.3.22)$$

where we have chosen to have $z=0$ at $t=0$ and z_1 is some constant. Notice that the vertical velocity, as the star passes through the plane, will be $z_1\omega$, so we can more usefully write the vertical displacement as

$$z = (v_z(0)/\omega) \sin \omega t \qquad (4.3.23)$$

Returning to the radial motion, we can see that the star could have a perfectly circular orbit only at the radius where the angular velocity is

$$\Omega(R_*) = \frac{L_z}{R_*^2} \qquad (4.3.24)$$

The real star will deviate from this perfect circular motion, but will only move a small distance from its so-called 'guiding centre', which does follow this motion. Setting $R = R_* + r$ (with $r \ll R_*$) and using the Taylor expansion again, we obtain

$$\frac{d^2 r}{dt^2} \simeq -r \left[\frac{\partial^2}{\partial R^2}\left(\Phi + \frac{L_z^2}{2R^2}\right)\right]_{R_*} \qquad (4.3.25)$$

But for circular motion

$$R^2 \Omega(R) = \frac{\partial \Phi}{\partial R} \qquad (4.3.26)$$

So this becomes

$$\begin{aligned} \frac{d^2 r}{dt^2} &= -r\left[\frac{d}{dR}(R\Omega^2) + \frac{3L_z^2}{R^4}\right]_{R_*} \\ &= -r\left[\frac{1}{R^3}\frac{d}{dR}(R^4\Omega^2)\right]_{R_*} \end{aligned} \qquad (4.3.27)$$

We can write this finally as

$$\frac{d^2 r}{dt^2} = -\kappa^2 r \qquad (4.3.28)$$

with

$$\kappa^2(R_*) = \left[\frac{1}{R^3} \frac{d}{dR} (R^4 \Omega^2) \right]_{R_*} \qquad (4.3.29)$$

If we take R_* to be the Solar position R_0, then we can use the definition of Oort's constant B to simplify this to

$$\kappa^2 = -4B\Omega \qquad (4.3.30)$$

Since $B < 0$ it is ensured that κ^2 is positive and we again have simple harmonic motion, now in the radial direction, with

$$r = R - R_* = r_1 \sin \kappa t$$

As before, we can rewrite this in terms of the star's outward velocity as it crosses $r = 0$ as

$$r = (v_R(0)/\kappa) \sin \kappa t \qquad (4.3.31)$$

The star therefore oscillates around the mean radius, and to conserve angular momentum also oscillates in azimuth relative to its guiding centre. It thus undergoes 'epicyclic motion' (as in Ptolomy's description of planetary motion), moving around a small loop with 'epicyclic frequency' $\kappa = (-4B\Omega)^{1/2}$ about its mean position (which itself revolves around the Galaxy, of course). Specifically, the azimuthal motion follows from

$$\frac{d\theta}{dt} = \frac{L_z}{R^2} = \Omega(R_*) R_*^2 (R_* + r)^2 \simeq \Omega(R_*) \left(1 - \frac{2r}{R_*} \right) \qquad (4.3.32)$$

Substituting for r from above and integrating, we obtain

$$\theta(t) = \Omega(R_*)t - \frac{2\Omega(R_*)}{R_*} \frac{v_R}{\kappa^2} \cos \kappa t \qquad (4.3.33)$$

Notice that the '$R\theta$' motion is $90°$ out of phase with radial motion and $2\Omega/\kappa$ times larger (hence elliptical epicycles, *unlike* Ptolomy's).

For a point mass $\Omega \propto R^{-3/2}$, so $\kappa = \Omega$ and the epicycle superimposed on the circular motion describes the overall elliptical orbit (with the Galactic Centre at one focus), just as for planetary motion. For a sphere of constant density $\Omega(R)$ is constant and $\kappa = 2\,\Omega$, so in fact the epicycles are now circles and the star makes two excursions per orbit, which therefore becomes an ellipse *centred* on the Galactic Centre. Near the Sun, $\kappa \simeq 1.4\,\Omega$ (i.e. between the two simple cases), so the stars make about 1.4 oscillations per orbit, which is therefore not closed. The epicyclic period is 170 Myr compared to the orbital period of 240 Myr.

4.3.3   The velocity ellipsoid

We can also compare the velocities of the stars in the radial and azimuthal directions. Relative to a frame rotating at angular speed $\Omega(R_0)$, the azimuthal speed of a star passing the Sun (i.e. with $R_* + r = R_0$),

$$v_\theta = R_0 \left(\frac{\mathrm{d}\theta}{\mathrm{d}t} - \Omega(R_0) \right) \simeq R_0 \left[\Omega(R_*) - 2r\frac{\Omega(R_*)}{R_*} - \Omega(R_0) \right]$$

$$\simeq -r \left[2\Omega(R_0) - R_0 \left(\frac{\mathrm{d}\Omega}{\mathrm{d}R} \right)_{R_0} \right] = 2Br \tag{4.3.34}$$

Averaging over all local stars

$$<v_\theta^2> \;=\; 4B^2 <r^2> \tag{4.3.35}$$

But from the epicyclic (harmonic) motion, the radial velocity component v_R must scale as κr, so

$$<v_\theta^2> \;=\; \frac{4B^2}{\kappa^2} <v_R^2> \tag{4.3.36}$$

Finally, using $\kappa^2 = -4B\Omega$ and $\Omega(R_0) = A - B$, we have

$$\frac{<v_\theta^2>}{<v_R^2>} \;=\; \frac{-B}{A - B} \tag{4.3.37}$$

Since B is negative this implies that $<v_\theta^2>$ is less than $<v_R^2>$. The observational ratio is about 2, consistent with $B \simeq -A$, as before. As there is also a z component to the velocity dispersion, we have in general what is called a 'velocity ellipsoid' with axes $<v_R^2>$, $<v_\theta^2>$ and $<v_z^2>$. Note, too, that because there is a higher density of stars nearer to the Galactic Centre than further out, typical stars seen near the Sun will have *positive* values of r, and thus negative values of v_θ. They will therefore orbit more slowly than the LSR. This effect is known as 'asymmetric drift'.

4.4   Stellar populations

The discs, and especially the arm regions, of a spiral galaxy are noticeably blue in a colour photograph, while the bulge is yellow. This suggests that the different regions contain different sorts of stars. Using observations made with the Mount Wilson 100″ telescope during the wartime blackout in California, Walter Baade observed that the brightest stars in the bulge of M31 were red giants, rather than the blue supergiants found in its disc. He therefore proposed that the stars in a spiral galaxy in fact comprise two separate populations. Population I stars live in the disc and arms, while Population II stars inhabit the bulge, the globular clusters and the diffuse spheroidal extension of the bulge, the halo. Although a simplification, this provides a useful starting point.

Studies of the colour–magnitude (or Hertzsprung–Russell) diagram for stars in globular clusters in our Galaxy reveal that they are generally old systems, with main-sequence turn-off points indicating ages of at least 10, and probably 12, Gyr. A useful confirmation of these ages lies in the presence of variable stars known as RR Lyraes (after their prototype). These are lowish mass post-main-sequence stars which are burning helium in their cores and which all have closely similar luminosities, around $50L_\odot$, making them useful distance indicators. Stars take at least 8 Gyr to reach this stage of their evolution.

Furthermore, the precise colours of the stars at a given magnitude tell us the metallicity of the system, as we discussed for ellipticals. Most Galactic globular clusters are very metal poor with [Fe/H] $\simeq -2$, though some reach [Fe/H] $\simeq -0.6$, that is metallicities 1/100 to 1/4 solar (cf. section 3.5). Studies of halo stars in the outer parts of the Galaxy imply similar ages and metallicities there, too. We can nowadays also study the globular cluster systems of other spiral galaxies, and even though we have to use the colours of the integrated light in this case, we again arrive at similar conclusions. In fact, recent work has shown that there are two populations of globular clusters in our Galaxy and some others. There are the outer 'halo' globular clusters, which share low metallicities with the halo stars, and a closer-in population, sometimes called 'disc' globular clusters (though they are now thought to be more associated with the bulge). The latter are the ones with higher metallicities and also have somewhat younger ages (though they are still several Gyr old).

Bulges turn out to be more complex than originally thought, even though the bulge stars are all old. The bulge in our Galaxy is best viewed in the near infra-red, where dust extinction in the intervening disc is minimised – the extinction in the optical B band is over 30 magnitudes towards the Galactic Centre. Such observations suggest that the bulge contains around 20% of the stellar mass of the Galaxy and that the bulge is actually somewhat bar shaped, with the nearer end at positive Galactic latitudes ($\ell > 0°$). This suggests that, seen from the outside, our Galaxy would probably have a classification on the de Vaucouleurs system around SABbc.

Bulges of external galaxies appear photometrically to be an inward continuation of the stellar halo, but the evidence from our Galaxy is that the halo and bulge differ in their make-up. Specifically, the metallicities of the bulge stars show a wide range and are typically around half solar, compared to around 1/100 solar or less in the halo.[4] Also, as we noted earlier, the bulge has a significant rotation, around $100\,\mathrm{km\,s^{-1}}$ and is probably rather flattened, while the halo is essentially non-rotating and probably nearly spherical.

---

[4] The recently discovered HE 0107-5420 holds the record as the metal-poorest halo star, with [Fe/H] $\simeq -5.3$, or 1/200 000 solar.

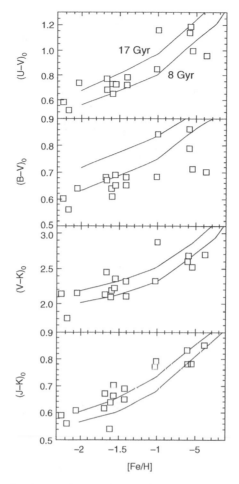

Figure 4.11 Variation of colour indices with metallicity for Galactic globular clusters. The solid lines are predictions of population synthesis models at the ages indicated (Reproduced with permission from Worthey, 1994).

In all cases we can identify the Population II stars with ones formed early in the histories of the galaxies. The universe is now believed to have a well-determined age close to 13.7 Gyr, so the galaxies cannot be older than about 13 Gyr. At the other extreme, consider 'open clusters' in the Galaxy, such as the well-known Pleiades. These are much less populous systems than globulars, containing between $\sim 10^2$ and $\sim 10^4$ stars. From their H–R diagrams we can see that open clusters still contain bright main-sequence stars, so must be quite young. The Pleiades stars are only about 100-Myr-old and very few open clusters have ages more than 1 Gyr. The ones that are that old are mostly situated in the outer parts of the Galaxy; in the inner parts, gravitational interactions lead to the dispersal of clusters on this timescale. The detailed positions of the stars in the H–R diagram also indicate metal abundances not very different from the Sun's. These are the classic characteristics of

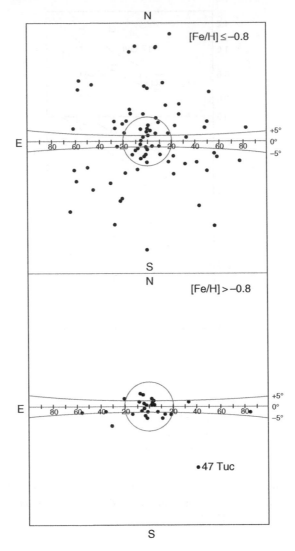

Figure 4.12   The distributions of metal-poor (top) and metal-rich globular clusters (Reproduced with permission from Zinn, 1985).

Population I stars. Note that this means that Population II stars formed before Population I stars! If there was any even earlier generation of stars, now not visible, they belong to Population III.

4.4.1   Colours

Since bulge stars are older and therefore redder on average than disc stars, the overall colour of spiral galaxies must change systematically with B/D ratio along

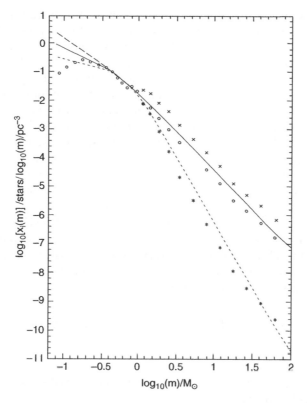

Figure 4.13   The derived present-day mass function (PDMF; short dashed line) and initial mass function (solid line). The PDMF is steeper because lower mass stars have been accumulating for a longer period (Reproduced with permission from Kroupa, Tout and Gilmore, 1993).

the Hubble sequence. Early type spirals are quite red, like S0s, say B–$V \sim 0.9$, while late types are quite blue, B–$V \sim 0.5$, with corresponding changes in optical-infrared colours, too. In terms of their spectra the earlier type spirals are again much like S0s, with spectra reminiscent of K-type stars, showing H and K lines, the G-band and so on. By the time we reach Scs we find much more energy at short wavelengths, dominated by the output from O and B stars, as well as many emission lines arising from regions of star formation.

4.4.2   The initial mass function

In more detail, the stellar population in terms of individual stars in any given region of a galaxy is determined both by the times the stars formed – the star formation history (SFH) and the distribution of masses of the stars which formed in any particular star formation episode, the IMF. Although the IMF may vary from its usual form in extreme environments or at very early cosmic times, it seems to be a reasonable approximation, observationally, to assume a universal IMF. A number of forms have been suggested, the most often used being the Salpeter mass function

$n(M) \propto M^{-2.35}$, as in section 3.4.2. However, this probably overestimates the number of low mass stars, so a better approximation may be a double power law, with slopes $\simeq -2.3$ for massive stars and $\simeq -1.3$ at low masses (below about half a solar mass). The overall range of stellar masses is often taken to be 0.1 to $100 M_\odot$.

Of course, what we actually observe in some region of a galaxy – for instance the Solar Neighbourhood – is not the mass function but the luminosity function. Looking just at the main-sequence stars, we can determine the LF in exactly the same way as we have discussed for galaxies. This then differs from the IMF in two ways. Firstly we have to translate between main-sequence luminosity and stellar mass via models of stellar structure, and secondly we need to allow for the fact that the high mass stars we see are only those which have formed in the last few million years, whereas the low mass stars have been accumulating since the Galaxy formed $\sim 10^{10}$ years ago. We can either assume that stars have formed at a uniform rate over the whole lifetime of the Galaxy, or we can attempt to determine the SFH and fold that into the calculation. The latter requires a measurement of either the detailed distribution of stars across the H–R diagram or precise observations of representative stars' colours and spectral features. Evidently, the present-day $(t = t_0)$ number density of main-sequence stars of mass M must be

$$N_0(M) = \int_{t_1(M)}^{t_0} b(M, t)\mathrm{d}t = n(M) \int_{t_1(M)}^{t_0} \Psi(t)\mathrm{d}t \qquad (4.4.1)$$

where b is called the stellar birthrate function and $\Psi(t)$ is the total star formation rate. The interval $t_0 - t_1(M)$ is the main-sequence lifetime of stars of mass M, which is given empirically (in years) by

$$\log \tau_{\mathrm{ms}}(M) \simeq 10.0 - 3.42 \log(M/M_\odot) + 0.88 \log (M/M_\odot)^2 \qquad (4.4.2)$$

(so roughly, for near solar mass stars, $\tau_{\mathrm{ms}} \simeq 10^{10} M^{-3.4}$ yr).

A simpler method, though limited by the relatively small numbers of stars present, is to study individual young clusters such as the Pleiades where we can assume that all the stars formed in a single star formation event. As for the analogous case for galaxies in clusters, this also has the advantage of better sampling the very faint objects, so we can extend the IMF right down to red dwarf and even brown dwarf stars.

A particular consequence of the form of the IMF is the rate of production of supernovae (SNe). For instance, for a Salpeter IMF, we expect about one star above $8 M_\odot$ – and hence able to produce a Type II (core collapse) SN – for every $135 M_\odot$ turned into stars.

4.5   The interstellar medium

Even in our Galaxy, the ISM began to be studied in detail much more recently than the stellar content. Recall, though, that Curtis had invoked the presence of interstellar dust to explain why many more spiral nebulae were observed in the regions away from

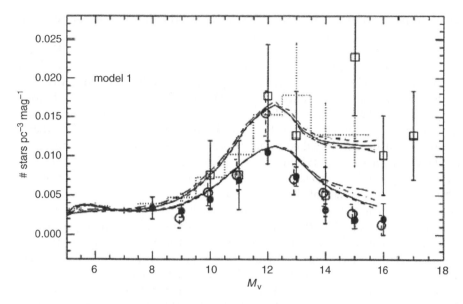

Figure 4.14 A compilation of luminosity functions of low luminosity stars near the Sun (Reproduced with permission from Kroupa, Tout and Gilmore, 1993).

the plane of the Galaxy (seen on the sky as the Milky Way). The obscuring effects of dust patches or clouds were obvious even in the earliest photographs of the Milky Way by E.E. Barnard, but it was R.J. Trumpler in the 1930s who first quantified the amounts of general interstellar dust. Trumpler compared the H–R diagrams for star clusters in the Galactic Plane with that for nearby stars to deduce a photometric distance to the clusters. He also measured the angular sizes of the clusters, and using his inferred distances calculated their physical sizes. These implied that clusters became systematically larger the further away they were. To avoid this implausible conclusion, he argued that the inferred distances were too large: the stars looked fainter not only because of the inverse square law but also because their light was dimmed by roughly a factor 2, or 0.7 magnitudes, per kpc of sightline through the disc (cf. section 2.6). His results also accounted for the then unexplained discrepancy between the size of the 'Kapteyn Universe' deduced from star counts and the size of Shapley's model of the Galaxy based on globular cluster distances.

4.5.1   HI

It was not until the 1950s that the next steps were taken. In 1944, Henk van de Hulst predicted that neutral hydrogen would emit in the radio waveband at 21.1 cm (1420 MHz). This radiation is due to what is called a hyperfine transition. Besides the quantum numbers representing the particular 'orbit' which the electron in a hydrogen atom occupies, the electron has a 'spin' quantum number. If this spin is in the same direction as that of the proton which it is orbiting, then the atom has a slightly different energy than if the two spins are in opposite directions. If the electron 'flips' from one spin state to the other then the tiny amount of excess energy ($\sim$6 $\times$ 10^{-6} eV) is released as a 21-cm photon.[5]

Radio astronomy had begun just a few years earlier with the work of Karl Jansky and Grote Reber, but after wartime developments in radio technology the 21-cm line emission from the Galaxy was detected by groups in the Netherlands, Australia and the USA, more or less simultaneously, in 1951. The 21-cm line has since become *the* key to studies of the ISM in our Galaxy and other spirals. Although the transition is quite unlikely – a given electron is likely to flip only once in 3.5×10^{14} s – there is so much neutral hydrogen in galaxies that it produces by far the most important radio line. The neutral hydrogen (HI) mass of our Galaxy, for instance, is about $6 \times 10^9 M_\odot$, corresponding to 10^{66} hydrogen atoms, 3×10^{51} of which will flip each second.

The relative importance of the ISM in galaxies is often quantified by the hydrogen mass to optical light ratio M_H/L_B in solar units, that is, in $M_\odot/L_{B\odot}$. For an early type spiral this may be less than 0.1, but this rises to about 0.8 for late types and for the class of spirals labelled 'HI rich' may exceed 1. For average spirals this then corresponds to a relative mass in gas, compared to stars, of order 0.1 to 0.2. The rare HI rich galaxies, where this figure rises to around a half, are frequently of rather low surface brightness, suggesting that they may have been inefficient at forming stars.

The other way to quantify the HI content is directly through the HI mass function or HIMF. The main difficulty here, though, is that until the late 1990s surveys, by necessity, targeted optically catalogued galaxies – generally large galaxies at known recession velocities – in order to 'tune' the radio observations to the correct range of frequencies. Galaxies with large HI contents but which are inconspicuous in the optical will therefore be under-represented.

In the last few years, though, technical advances with 'multi-beam' receivers have made possible 'blind' (i.e. non-targeted) searches of large areas of sky, using multiple simultaneous observations. The HI Parkes All-Sky Survey (HIPASS) and its northern hemisphere counterpart from Jodrell Bank (HIJASS) have surveyed the whole local volume of space out to recession velocity $\simeq$10 000 km s^{-1}, or $D \simeq 140$ Mpc. The detection limit is around $M_{HI}/M_\odot = 10^6 D^2$, meaning that they could detect $10^6 M_\odot$ of HI at $D = 1$ Mpc, but need $10^{10} M_\odot$ for a detection at $D = 100$ Mpc.

The observed distribution of HI masses spans M_{HI} from $10^{6.8}$ to $10^{10.6} M_\odot$ and peaks at about $10^{9.5} M_\odot$. However, allowing for the smaller volumes surveyed for the smaller masses produces a HIMF with a rather similar shape to the optical LF. The low mass end is still not very well constrained but probably has a power law slope around -1.3 and, again like the optical LF, is steeper for later type galaxies. One rather different selection effect than those encountered in the optical occurs because radio detections need to be over several frequency channels in order to distinguish them from interference. Thus narrow lines will not be detected. This will tend to reduce the number of low mass galaxies, with low rotation velocities, and also biases the survey against nearly face-on galaxies (which will have only a small radial component of velocity from their rotation).

---

[5] Fortunately, due to wartime conditions in the Netherlands, van de Hulst had been unaware of an earlier calculation which had 'proved' that hyperfine transition lines would be unobservable.

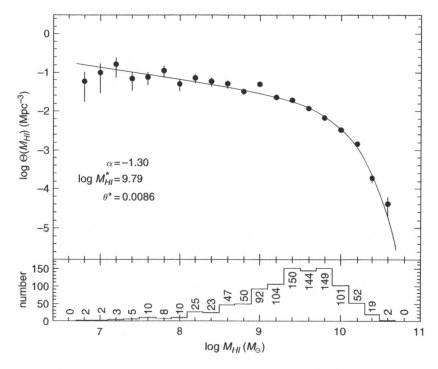

Figure 4.15 The HI mass function from the HIPASS survey. The lower panel shows the number of galaxies actually observed in each bin before correction for the observable volumes (Reproduced with permission from Zwaan *et al.*, 2003).

4.5.2 Other constituents of the ISM

Not all the hydrogen in a galaxy need be neutral atoms; we also expect contributions from ionised hydrogen (HII) and molecular hydrogen (H_2). (As in the Sun and the rest of the Universe, almost all of the remaining 25% of the mass of the ISM which is not hydrogen in some form is helium.) The ionised gas is detected mainly via the emission lines it produces, especially the recombination lines resulting when electrons drop back into the lower energy levels. HII regions – spheres of ionised gas around hot stars – have particularly characteristic hydrogen Balmer line spectra ($H\alpha$, $H\beta$, etc.) as seen in Figure 4.16.

Molecules, on the other hand, radiate via rotational and vibrational transitions of the whole molecule. H_2 should be by far the most abundant molecule, but is an inconvenient one to observe since it is symmetrical and therefore has no dipole moment. Only weak quadrupolar transitions are therefore possible and there are no useful optical or radio emission lines. Instead, observers tend to use measurements of the much rarer but much more amenable CO molecule, though this then leads to the added complication of determining the correct translation factor between the mass in CO and that in H_2. The value (or variety of values) of this factor has been a topic of considerable controversy over the years, with the

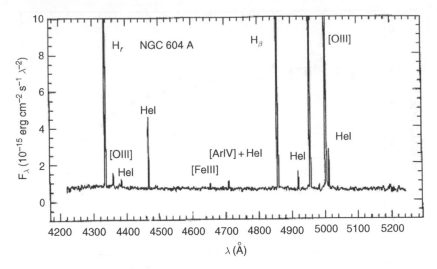

Figure 4.16 The characteristic emission line spectrum of an HII region (Reproduced with permission from Diaz *et al.*, 1987).

likelihood that it will be different for metal-rich and metal-poor galaxies, for instance. A reasonable conversion, at least for galaxies like ours, is probably

$$N_{H_2} \simeq 1.2 \times 10^4 N_{CO} \tag{4.5.1}$$

where the CO (strictly the isotope ^{12}CO) is measured via the rotational transitions at 1.3 and 2.6 mm. These lines are particularly strong for molecular densities $\sim 10^9 - 10^{10} \, \text{m}^{-3}$, and the transitions are excited even at very cold interstellar temperatures of 10 or 20 K. Our Galaxy appears to have quite a large molecular mass, around $3 \times 10^9 \, M_\odot$. Lines from the isotope ^{13}CO or other rarer molecules with large dipole moments (essentially the distance between the centre of mass and the centre of the charge distribution), like NH_3 and HCN, can be used as diagnostics of higher densities; the centres of Giant Molecular Clouds (GMCs) can reach $\sim 10^{12}$ molecules per m^3. Interstellar chemistry, the study of the production of such molecules in space, is a whole field of study in its own right, and we will not enter into its complexities here.

4.6 Neutral gas

If we generalise our earlier result, from section 2.6, for the absorption of radiation by an intervening medium, we can see how much 21-cm radiation we should receive from a cloud of HI. The cloud will both emit and absorb radiation along our line of sight, so in passing through some thickness ds, the resultant intensity I must vary according to

$$\frac{\mathrm{d}I}{\mathrm{d}s} = \eta - \kappa I \tag{4.6.1}$$

where η is the emissivity and κ, as before, is the absorption coefficient. If we define the 'source function' S to be the ratio of emission to absorption, η/κ, and recall the definition of optical depth

$$\tau = -\int \kappa \mathrm{d}s \tag{4.6.2}$$

then the radiation transfer equation becomes

$$\frac{\mathrm{d}I}{\mathrm{d}\tau} = I - S \tag{4.6.3}$$

If S is constant across a medium of total optical depth τ_ν at frequency ν, then, neglecting stimulated emission, the observed intensity outside the cloud will be simply

$$I_\nu = S(1 - e^{-\tau_\nu}) \tag{4.6.4}$$

Collisional excitation bumps virtually three-quarters of the electrons into the higher of the two spin states, the ratio appropriate for perfect thermodynamic equilibrium, since the two hyperfine energy levels are separated by an energy difference $h\nu$ equivalent to $k_B T$ for T of only 0.07 K, very much less than typical HI temperatures of 100 K. Hence the source function is the Planck or blackbody function

$$B_\nu(T_s) = \frac{2h\nu^3/c^2}{e^{h\nu/k_B T_s} - 1} \tag{4.6.5}$$

where the 'spin temperature' T_s is virtually indistinguishable from the kinetic temperature of the atoms.

Now, since $h\nu \ll k_B T_s$ we can take the Rayleigh–Jeans limit of the Planck function

$$B_\nu(T_s) = \frac{2k_B T_s \nu^2}{c^2} \tag{4.6.6}$$

which implies

$$I_\nu = \frac{2k_B T_s \nu^2}{c^2}(1 - e^{-\tau_\nu}) \tag{4.6.7}$$

By analogy with the form of the Planck function, radio astronomers often define a 'brightness temperature' T_b such that

$$I_\nu = \frac{2k_B T_b \nu^2}{c^2} \tag{4.6.8}$$

so in these terms

$$T_b = T_s(1 - e^{-\tau_\nu}) \tag{4.6.9}$$

4.6.1   HI observations

For an optically thick source (large τ_ν) $T_b = T_s$, so from equation (4.6.9) we can say nothing about the total amount of hydrogen in the line of sight. However, when τ_ν is small $T_b \simeq \tau_\nu T_s$ so the brightness temperature is proportional to the optical depth and hence the HI column density N_H. Numerically, the optical depth at frequency ν turns out to be

$$\tau_\nu = \frac{CN_H\phi_\nu}{T_s} \tag{4.6.10}$$

with $C = 2.57 \times 10^{-15}$ for T in kelvin, ν in Hz and column densities in their traditionally quoted units of atoms per square cm. ϕ_ν here describes how 'spread out' the emission is in frequency, i.e. $\phi_\nu d\nu$ is the fraction of atoms which can emit or absorb photons of frequencies ν to $\nu + d\nu$.

Inverting this and integrating over all relevant frequencies we obtain

$$N_H = 3.88 \times 10^{14} \int T_s \tau_\nu d\nu \tag{4.6.11}$$

Since for a given line the observed frequency is also equivalent to a Doppler velocity ν (in km s^{-1}), we can alternatively write this as

$$N_H = 1.82 \times 10^{18} \int T_b(\nu) d\nu \tag{4.6.12}$$

If we are looking at an external galaxy at distance D, we will need to integrate over its whole apparent area $A = D^2\Omega$, where Ω is the solid angle it subtends on the sky. The total number of hydrogen atoms is then

$$N_T = 1.82 \times 10^{18} D^2 \int d\nu \int T_b d\Omega \tag{4.6.13}$$

If D is measured in Mpc and we revert from brightness temperature to observed flux F – which we will measure in units of janskys, where $1\,\mathrm{Jy} = 10^{-26}\,\mathrm{W\,m^{-2}\,Hz^{-1}}$ – then we finally obtain the total mass (for the optically thin case, remember)

$$M_{HI} = 2.36 \times 10^5 D^2 \int F_\nu d\nu \, M_\odot \tag{4.6.14}$$

Despite the huge amounts of HI in a galaxy, the 21-cm line is still generally optically thin since, even ignoring the rotation of the disc, the turbulent velocities in the gas will spread the line out with a Doppler width of around $10\,\mathrm{km\,s^{-1}}$, corresponding to a frequency width of $\delta\nu \simeq 5 \times 10^4$ Hz. For a hydrogen density of 0.5 atoms per cm^3, say, we accrue a column density $N_H \simeq 10^{21}$ cm^{-2} along a pathlength of 2×10^{21} cm or 700 pc. If we assume that ϕ_ν is a gaussian of width $\delta\nu$ then ϕ_ν takes the value $1/(\sqrt{\pi}\delta\nu) \simeq 10^{-5}$ at the centre of the line. Thus with $T_s \simeq 125$ K our equation above gives $\tau_\nu \simeq 0.2$.

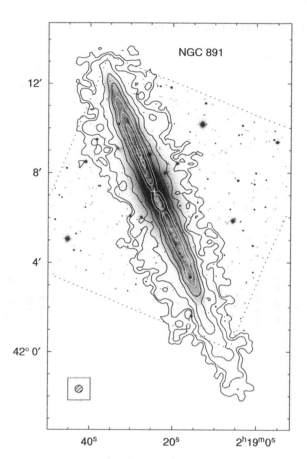

Figure 4.17 The HI layer in the edge-on galaxy NGC 891 (Reproduced with permission from Swaters, Sancisi and van der Hulst, 1997).

Thus along most sight lines through a galaxy (or through our Galaxy, unless we are looking directly towards the centre), HI is optically thin to its own radiation and we can use the intensity to infer the column density and total hydrogen mass as above.

Since each photon has energy $h\nu$, the rate at which photons will be received, in some bandwidth $\Delta\nu$, from a source of angular size Ω (or with a radio telescope of beam size or resolution Ω) will be

$$n_\nu = 2\tau_\nu \frac{k_B T_s}{h\nu} \frac{\nu^2}{c^2} \Delta\nu \, \Omega \, \mathrm{m}^{-2} \, \mathrm{s}^{-1} \tag{4.6.15}$$

A 100-m dish with a beam size around 1 square arc minute and a receiver with resolution $\simeq 1\,\mathrm{kHz}$ observing a cloud with $T_s \sim 100\,\mathrm{K}$ should therefore collect about 100 21-cm line photons per second per channel. These will have a total energy of $10^{-3}\,\mathrm{eV}$.

4.6.2 The HI and HII distributions

The atomic (and ionised) ISM in most spiral galaxies lies in a rather thin layer, with a scale height typically around 150 pc. Its radial extent is often much greater than that of the bulk of the stars, however, and the gas layer is frequently seen to 'flare' out far from the centre. It may also 'warp' away from the central plane as defined by the disc of stars. Both these effects are seen for our Galaxy, where the extent of the HI gas is at least 25 kpc.

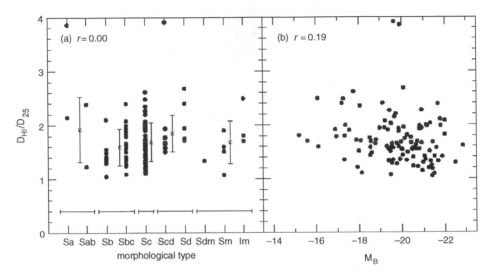

Figure 4.18 The distribution of the ratio of HI disc size to optical size, as a function of spiral type and luminosity (Reproduced with permission from Broeils and Rhee, 1997).

Many spirals, including our own, show their peak HI column density some distance away from the galaxy centre. M31 is even more extreme, with basically a hole in the distribution inside a few kpc. Overall, though, the HI density varies relatively little across most discs. Peak values are usually less than or about 10^{21} atoms per cm^2, while the radius at the 10^{20} atoms per cm^2 level is on average twice the optical R_{25}. This means that there is a drop of only a factor $\sim$10 in column density while the optical disc light falls by about 6 magnitudes or a factor 250.

Unsurprisingly, the HI size correlates with the optical size, while the HI mass correlates with both, indicating similar mean HI mass densities in most spirals. This mean value is probably determined by optical depth effects. A layer of column density above about $4 \times 10^{20}\,cm^{-2}$ ($3M_\odot\,pc^{-2}$) is essentially optically thick to all UV photons which could dissociate H_2 molecules. Thus we might expect that denser layers will self-shield their inner parts, allowing molecules to form at the expense of neutral atoms. This is consistent with the much clumpier distribution seen in molecular gas compared to HI, since the H_2 will occur only where the overall gas density is particularly high.

The molecular gas, which occurs mostly in distinct clouds, including the very massive GMCs with densities above $10^9\,\mathrm{m}^{-3}$ and $T \sim 20\,\mathrm{K}$, is thus even more tightly compressed into the plane than the HI, with a half-thickness less than 100 pc. Also contrary to the HI, the molecular gas is generally fairly close to the centres of spiral galaxies, though again it avoids the centres themselves in many cases, piling up in rings at radii of a few kpc. Our Galaxy has a particularly high molecular density at about 4 kpc from the centre.

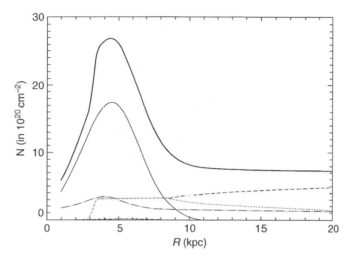

Figure 4.19   The radial distribution of cool gas in the Galaxy (compiled by Ferriere, 1998). The thin solid line shows the contribution from molecules, the dotted line that from the Cold Neutral Medium, the dashed line the Warm Neutral Medium and the dot dashed line the Warm Ionised Medium (WIM).

By coincidence or otherwise, with typical gas to dust ratios, galaxy discs become opaque to optical photons ($A_B \sim 1$) at a similar column density as for molecular shielding, of the order of 10^{21} atoms per cm^2, corresponding to a dust column $\sim 0.1 M_\odot\,\mathrm{pc}^{-2}$, since observationally

$$N_H \simeq 2 \times 10^{21} A_B\,\mathrm{cm}^{-2} \tag{4.6.16}$$

Conversely, in the outer parts of spiral discs there is insufficient column density of hydrogen even to absorb all the ionising photons in general intergalactic space (emitted by the overall population of galaxies and quasars). Thus below a column density $\sim 10^{19}\,\mathrm{cm}^{-2}$, we might expect the HI to drop sharply as any hydrogen present will be ionised.

Large halos of gas around galaxies can be detected by their absorption of light from distant quasars behind them, giving rise to what are called QSO absorption line systems (QSOALS). Optical spectra of quasars typically show the lines of the MgII doublet at 280 nm or, at higher (absorber) redshifts, the UV lines of CIV at 155 nm. The large number of QSOALS seen per unit redshift interval (i.e. over a given distance range), coupled with the known number density of largish galaxies seems

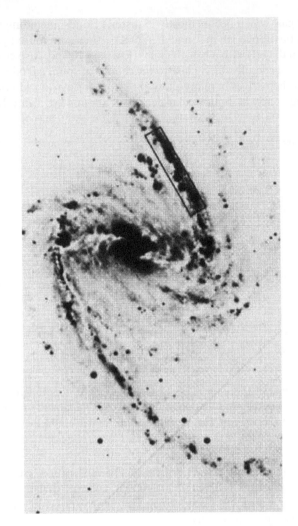

Figure 4.20   An AAT image of the galaxy NGC 1365, emphasising the strings of HII regions along its arms (Reproduced with permission from Roy and Walsh, 1988).

to imply that the absorbers should have extents of 2 to 4 Holmberg radii, or perhaps 40–80 kpc. These same absorption lines, when produced by ionised gas in our Galaxy, can be observed in the spectra of distant stars by using far UV observations from satellites such as the International Ultraviolet Explorer (IUE).

4.7   Ionised gas

Ionised hydrogen is seen not only in the outer extremities but also in the more central regions of galaxies. Luminous young stars, or clumps of them in star forming regions, emit copious ionising photons, so large parts of the interstellar medium are in the form of HII. Most prominent are the discrete HII regions ionised by a single star or tight grouping of them. In particular, if we have a hot star (surface temperature above 20 000 K) inside a hydrogen cloud of reasonably uniform density, then all the gas in a roughly spherical region around the star will be ionised. This region is known as a Strömgren Sphere.

In equilibrium, the ionisation rate must equal the rate at which electrons recombine with protons, emitting the characteristic line spectra. If the star is emitting N_{UV} ionising photons per second, this will be sufficient to ionise all the atoms in a volume of radius r_s such that

$$N_{UV} = \frac{4\pi}{3} r_s^3 n_e n_H \alpha_2$$ (4.7.1)

i.e.,

$$r_s = \left(\frac{3 N_{UV}}{4\pi n_e n_H \alpha_2} \right)^{1/3}$$ (4.7.2)

where n_e and n_H are the electron and hydrogen densities, respectively, and α_2 is the recombination coefficient for electrons *not* falling directly to the ground state. (Those that do, produce another ionising photon so have no net effect.) The captures to the $n = 2$ state or above result in a further non-ionizing photon as they drop to the ground state, most of them eventually producing the characteristic Balmer lines, especially the 656.3-nm $H\alpha$ line which gives HII regions and other ionised gas clouds their characteristic red or pink hue in colour photographs. Essentially, for every ionising photon absorbed, about one $H\alpha$ photon is emitted, so we can use $H\alpha$ emission to measure the ionising radiation. For an O5 star, $N_{UV} \simeq 5 \times 10^{49}$ photons per second, and for a typical cloud $n_e \simeq n_H \simeq 10^9 \, \text{m}^{-3}$, while at around 8000 K $\alpha_2 \simeq 3 \times 10^{-19} \, \text{m}^3 \, \text{s}^{-1}$. Thus we obtain a Strömgren radius of about 1 pc. In general HII regions are quite sharply 'ionisation bounded', because we run out of ionising photons (rather than atoms to ionise, the density bounded case).

HII regions can also be observed in the radio via their 'free–free' emission, caused when an electron loses energy by being deflected, or scattered, by a proton, but not captured. In this case the emission is characterised by the mean energy of the electrons (hence it is 'thermal' radiation) and again, as for the $H\alpha$, its intensity depends on the square of the electron density through the 'emission measure'

$$EM = \int n_e^2 \text{d}s$$ (4.7.3)

where we are integrating along the sightline through the cloud. Conventionally *EM* is quoted in units of cm^{-6} pc. Since the change of energy in a scattering can take any value, the free–free emission leads to radio continuum emission, with a characteristic power law spectrum $F_\nu \propto \nu^{-\alpha}$ with $\alpha \simeq 0.1$, except at low frequencies where it turns over to ν^2 (as for black body radiation) due to 'self-absorption', i.e. the medium becomes optically thick to its own radiation.

4.8 ISM structure

All in all then, the ISM is quite complex, exibiting a multi-phase structure, mostly in approximate pressure equilibrium. It contains cool dense material, including the molecular gas, at $T < 100\,\text{K}$ and $n > 10^7$ atoms per m^3, a mixture of neutral and ionised warm interstellar material (WIM) at a few thousand K and $n \sim 10^5\,\text{m}^{-3}$, and hot ionised plasma at $T \sim 10^6\,\text{K}$ and $n \sim 10^3\,\text{m}^{-3}$. Most of the gas *mass* is in the cool component, but the hot diffuse phase, with $\sim 10^{-4}$ times the density, fills most of the volume. The WIM extends to heights around 1–2 kpc above the plane and the hot plasma forms an extensive halo around the warm clouds. The Sun lies in a Local Bubble of hot gas about 300 pc across, shown up by ROSAT X-ray and extreme UV observations. Inside the bubble, the density is mostly less than $2 \times 10^4\,\text{m}^{-3}$ and the temperature is around $10^6\,\text{K}$. On a smaller scale there is the mostly neutral ($T \sim 7000\,\text{K}$) Local Interstellar Cloud, around 8 pc across, with the Sun near one edge.

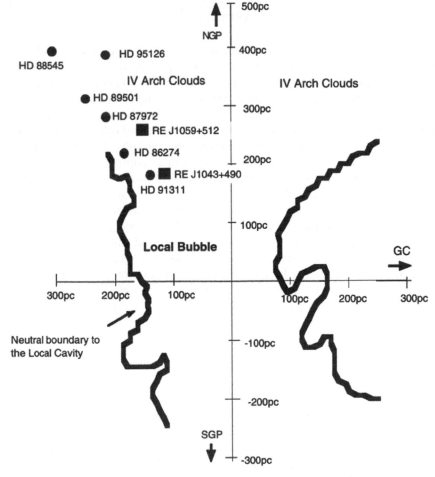

Figure 4.21 The structure of the local interstellar medium around the Sun (Reproduced with permission from Welsh, Sallmen and Lallement, 2004).

The different phases are in continual motion. Cool gas can be heated by supernovae explosions, for example, and the hotter gas will bubble up out of the plane into the halo – making a so-called 'galactic fountain' – before cooling and falling back in. Another, possibly related, component at large distances from the plane is the population of High Velocity Clouds of HI (HVCs) apparently falling into the Galaxy at high speeds $> 100\,\mathrm{km\,s^{-1}}$, though their origin and distances are still not firmly established.

4.8.1 Magnetic fields

The ISM is permeated by a magnetic field B of a few $10^{-10}\,\mathrm{T}$.[6] In fact, there are two components, one an ordered field aligned along the spiral arms, the other a tangled or random field. The energy density in these fields, $B^2/2\mu_0$, where μ_0 is the magnetic permeability ($4\pi \times 10^{-7}$ in standard units), is usually close to the thermal energy in the gas, $3/2\,nk_\mathrm{B}T$, typically a few $\times 10^5\,\mathrm{eV/m^3}$. This may suggest some sort of self-regulation or feedback loop. (This energy density also happens to be close to that in the CMB radiation field, $2.6 \times 10^5\,\mathrm{eV/m^3}$.) The close interaction of the field with the ISM plasma (charged particles find it hard to move *across* magnetic field lines) then explains the orientation of the field, since the gas swept up into the denser arms will carry 'frozen in' magnetic field with it.

Electrons accelerated by supernovae up to relativistic velocities (and energies typically of order GeV) will loop around the magnetic field lines and thereby radiate, primarily in the radio region, via the synchrotron process which we consider in more detail in Chapter 6. Since the electrons are moving at relativistic speeds around the field lines, the radiation they emit is 'beamed', that is the emitted photons are concentrated in the direction ahead of the electron.

If the electrons' energy distribution can be written as a power law

$$n(E) = n_0 E^{-\gamma} \tag{4.8.1}$$

(observations show this to be a good approximation with $\gamma \simeq 2.6$), then the brightness temperature of the radiation can be shown to be

$$T_b \propto n_0 B_\perp^{(\gamma+1)/2} \nu^{-(3+\gamma)/2} \propto B_\perp^{1.8} \nu^{-2.8} \tag{4.8.2}$$

where $B_\perp$ is the field component perpendicular to the path of the particle and therefore, because of the forward beaming, perpendicular to the observer's line of sight. Since the intensity $I(\nu) \propto T_b \nu^2$, this then implies a characteristic power law spectrum over a wide range of radio frequencies

$$I(\nu) \propto \nu^{-\alpha}$$

with $\alpha \simeq 0.8$.

---

[6] This is often given in the older unit of microgauss (μG), where $1\,\mathrm{G} = 10^{-4}\,\mathrm{T}$.

In general, the radio continuum radiation we see from spiral galaxies, say at 408 MHz, is a combination of this synchrotron radiation and thermal emission due to free-free radiation from the hot ionised gas, particularly that associated with the HII regions. In our Galaxy we may use the detailed distribution across the sky of radio flux at different observed frequencies to separate the two components, since they have different spectral slopes. We can therefore map out the synchrotron emisssion and thereby determine the distributions of relativistic electrons and magnetic fields in the Galaxy.

Other routes to determining the magnetic field include the 'Faraday rotation' of the plane of polarisation of any linearly polarised radiation (of wavelength λ) from background sources, such as pulsars. This depends on the product of the magnetic field component along the line of sight, $B_\parallel$, the electron density n_e and the path length Δs through the ISM as

$$\Delta\Theta = \text{RM}\,\lambda^2 \tag{4.8.3}$$

where the 'rotation measure'

$$\text{RM} = \frac{e^3}{2\pi m_e^4 c^4} n_e B_\parallel \Delta s = 8.12 \times 10^3 \; n_e B_\parallel \Delta s \; \text{m}^{-2} \tag{4.8.4}$$

for n_e in m^{-3}, B in tesla and Δs in parsecs. By determining the slope of $\Delta\Theta$ against λ^2, we can find the product $n_e B_\parallel \Delta s$.

To separate out the value of the magnetic field itself, we can use the 'dispersion measure'. This measures the delay between pulses' arrival times as a function of

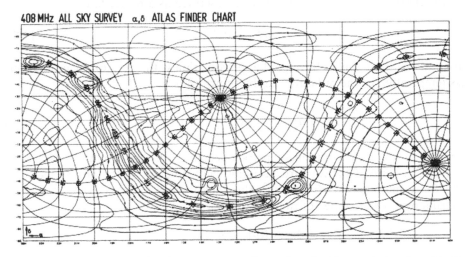

Figure 4.22 Radio map of the sky at 408 MHz (Reproduced with permission from Haslam *et al.*, 1982). The densely packed contours follow the line of the Galactic plane.

frequency, caused by the slower propagation of lower frequency waves through a plasma. This delay is given by

$$\Delta t = \text{DM } \lambda^2 \tag{4.8.5}$$

where

$$\text{DM} = \frac{e^2 \mu_0}{8\pi^2 m_e c^4} n_e \Delta s = 4.61 \times 10^{-8} n_e \Delta s \text{ m}^{-2}\text{s} \tag{4.8.6}$$

Thus the magnetic field can be obtained from the ratio RM/DM and values of a few μG (10^{-10} T) are again obtained.

4.8.2 The radio LF

For external spirals (and irregulars) we can determine a radio LF in exactly the same way as for the optical LF. The radio LF is often taken to be a power law (of slope $\simeq -3/2$) with a gaussian cut-off, for instance

$$\log \phi(L_{22}) = \text{constant} - 1.5 \log L_{22} - 0.28(\log L_{22})^2 \tag{4.8.7}$$

where L_{22} is the luminosity at 1.4 GHz in units of 10^{22} W Hz^{-1}. Note that converting between the different units implies that the radio luminosity is related to the observed flux F in Jy from a source at distance D Mpc via

$$\log L_{22} = \log F - 1.92 + 2 \log D \tag{4.8.8}$$

If we integrate over all radio powers, we then obtain a radio output per unit volume $\sim 2 \times 10^{19}$ W Hz^{-1} Mpc^{-3} at 1.4 GHz.

4.8.3 Cosmic rays

The electrons which produce the synchrotron radiation are one component of 'cosmic rays', high energy particles – also including atomic nuclei – which can be observed at Earth when they impinge on the atmosphere. The constituents of the cosmic ray flux, particularly the fraction of light element nuclei lithium, beryllium and boron, tell us how far the cosmic rays have travelled, or rather how much ISM they have travelled through, since the Li, Be and B nuclei are formed by 'spallation', when a larger nucleus collides with a particle in the ISM. The numbers of the various nuclei compared to Solar System abundances imply that the mean density in the ISM is $\sim 10^5$–10^6 atoms per m^3 and that cosmic ray ages are a few Myr.

Notice that since they are travelling at close to the speed of light, the cosmic rays will have travelled ∼1 Mpc in that time, requiring that they are somehow trapped in the Galaxy. This is possible because, like the electrons, their paths are curved by the Galactic magnetic field. The radius of curvature of their motion, due to the magnetic force $ZeBv$ is

$$r \simeq \frac{p}{ZeB} \qquad (4.8.9)$$

where p is the relativistic momentum and Ze the nuclear charge. The factor pc/Ze is known as the cosmic rays' 'rigidity': if it is high, i.e. at large energies per nucleon, the cosmic ray paths are only slightly curved. Cosmic rays with energies $\sim 10^{16}$ eV have r of order parsecs, but for those at 10^{19} eV it is several kpc. Cosmic rays of the highest energies seen, around 10^{20} eV, may originate outside our Galaxy.

The same cosmic ray particles are also responsible for the flux of gamma rays which emanates from the Galaxy. (The Galaxy has a γ-ray luminosity of about 10^{32} W, similar to that in X-rays.) γ-rays with energies ~ 100 MeV are produced as bremsstrahlung radiation when electrons of these energies scatter off the interstellar gas, as well as from the decay of pions produced by the interaction of cosmic ray nuclei of energies a few GeV with the ISM. The γ-ray emissivity is therefore proportional to the product of the cosmic ray and ISM densities, $\rho_{cr}\rho_{gas}$, again providing a measure of the cosmic ray density.

4.9   Dust

We have already mentioned the obscuring effects of interstellar dust in the disc of our Galaxy. Although its pervasive nature was not discerned for some time, individual dusty nebulae such as the Coal Sack, obscuring the stars behind them, are easily seen as dark patches within the Milky Way, even with the naked eye. On a slightly smaller scale, the famous Horsehead Nebula in Orion is a dark cloud superposed on a bright nebula, suggesting a connection between dust and star forming regions. Even smaller dark clouds, known as Bok Globules, are seen. Dust can, though, also result in bright patches. Reflection nebulae are seen via scattering of light from nearby or embedded bright stars and, exactly as for scattered sunlight in our atmosphere, appear blue in colour.

More generally, throughout the disc of the Galaxy, about 1% of the ISM mass is in dust, and the corresponding density n_g is about 10^{-6} grains per m³. The optical depth along a sightline of length s through the disc is then approximately $\tau_\lambda = \sigma_g n_g s$, where $\sigma_g = \pi a_g^2 Q_\lambda$ is the effective cross-section of a grain of radius a_g. Here Q_λ is the relative extinction coefficient, which measures how efficiently a grain removes photons of wavelength λ. In magnitudes, the extinction becomes

$$A_\lambda = 2.5 \log e\, \tau_\lambda = 1.086\tau_\lambda \qquad (4.9.1)$$

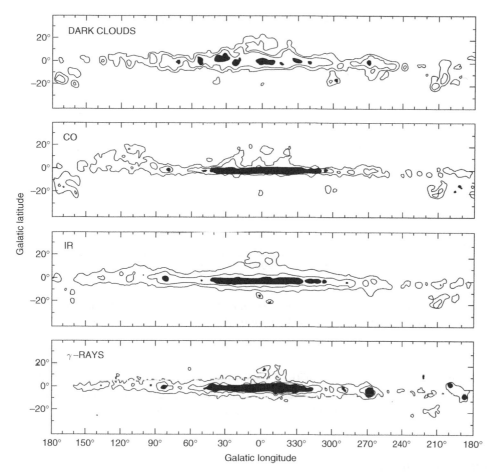

Figure 4.23 The distribution of dark clouds, molecular gas, infra-red emission and gamma-ray emission along the Galactic Plane (Reproduced with permission from Dame *et al.*, 1987).

The way in which extinction varies with wavelength, called the extinction curve, gives clues as to the make-up of the interstellar grains. The Galactic curve can be represented roughly by $\tau \propto \lambda^{-1}$, but shows, for instance, a bump at about 220 nm due to the presence of very small particles, around 0.01 microns in size, probably made of graphite. The UV extinction in general is thought to be produced by even smaller silicate grains, while the visible part of the curve results from, perhaps icy, grains around 0.1 microns across. There is also a contribution from PAHs (that is, polycyclic aromatic hydrocarbons), consisting of just 20 to 100 carbon atoms and with a size around 1 nm, as seen in the far infra-red emitting 'cirrus' clouds. The grains are probably formed in the atmospheres of cool supergiant stars and in GMCs. Given the respective masses and the overall gas-to-dust ratio, there must be of order 1 dust grain for every 10^{12} atoms in the ISM. The extinction curves for the LMC and SMC are somewhat different to the Galactic one, presumably as a

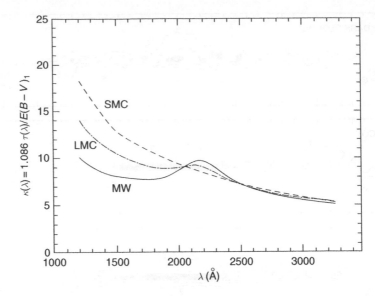

Figure 4.24 Interstellar extinction curves for the Galaxy and the Magellanic Clouds (compiled by Calzetti, Kinney and Storchi-Bergmann, 1994).

result of the lower metallicity of the ISM in these galaxies. Since different elements are more or less likely to be 'depleted' on to grains, some gas-phase ISM abundances differ substantially from 'standard' cosmic (or solar) values.

4.9.1   Reddening

The variation of τ with wavelength also gives rise to 'interstellar reddening', since longer wavelengths are less effected. We define the reddening to be, for example, using the B and V bands,

$$E(B - V) = A_B - A_V \tag{4.9.2}$$

A star of true colour $(B - V)_0$ will then be observed to have an apparent colour

$$(B - V) = (B - V)_0 + E(B - V) \tag{4.9.3}$$

Typically, it may be easier to determine the reddening than the actual absorption (for instance, if we know the spectral type of the star and hence its intrinsic colour). To get at the total absorption and thereby the true (dust free) apparent magnitude, we then need the ratio of total to 'selective' extinction

$$R_V = \frac{A_V}{E(B-V)} \qquad (4.9.4)$$

This is often asssumed to take the value 3.1 in the Milky Way, but is probably variable from place to place.[7] Extending this from Galactic stars, clearly observations of external galaxies, seen through the dust layer in the Galaxy, will be effected in the same way; so we need to take care of both overall dimming and the reddening of their light when determining their absolute magnitudes and colours.

4.9.2   Optical depths

Now consider the effects of dust inside other galaxies themselves. In the absence of dust, we will see the same total flux (apparent magnitude) irrespective of the inclination of the galaxy to our line of sight. However, the apparent surface brightness of an inclined-disc galaxy will change. As we tilt the galaxy relative to our line of sight, the path length through the stellar distribution must increase, thereby increasing the surface intensity. For thin discs, the inclination $\cos i \simeq b/a$, where a and b are as usual the lengths of the major and minor axes of the elliptical image. Thus the surface brightness in magnitudes at any point on the major axis will be

$$\mu(R) = \mu_0(R) - 2.5 \log(a/b) \qquad (4.9.5)$$

Alternatively, if we calculate the mean surface brightness $<\mu>$ from the apparent magnitude m simply as $m + 5 \log a$, then to the first order this should be unchanged by inclination since the observed total flux and size are still the same. (In fact, each should rise slightly as the isophotal limit will be slightly further out along the major axis for inclined systems.)

However, we know from observation within our Galaxy that dust obscuration increases as we look through longer path lengths at lower Galactic latitudes (section 2.7.1). We should therefore expect a similar increase in absorption if we look at inclined external spiral galaxies like, say, M31. The classic study of this was by Erik Holmberg in 1958. By fitting a function of the form

$$<\mu> = <\mu(0)> + A_B\left(\frac{a}{b} - 1\right) \qquad (4.9.6)$$

he obtained $A_B \simeq 0.3\text{--}0.4$. Note that here $\mu(0)$ refers to the surface brightness the disc would have it was face-on, not with no dust present. In using this, we are again assuming that the principal effect is on the total flux, not the observed isophotal major axis length.

---

[7] When the elderly Walter Baade was asked whether, given the chance to relive his life, he would still want to be an astronomer, he is supposed to have replied 'Only if given a guarantee that the ratio of total to selective absorption is the same everywhere!'.

Such a small value of A_B implies that galaxy discs are pretty much optically thin (recall section 2.6). However, we have implicitly assumed that the dust acts as a 'screen' in front of the stars (as it really does when we look out from our position in the Galaxy). A more realistic case is with the absorbing dust uniformly mixed in with the stars, or perhaps even in a thinner layer in the middle of the stars. These 'slab' and 'sandwich' models generate much less obscuration (and hence change in m or $<\mu>$) than does a screen. For instance, in the extreme case of an optically thick – even completely opaque – layer very close to the central plane of the disc, half of the stars are still visible and half are obscured *regardless* of the inclination.

In more detail, consider a simple uniformly mixed slab of physical thickness X containing stars of volume emissivity ε and dust of total optical depth τ (i.e. optical depth per unit path length τ/X). In the face-on case the observed intensity from a volume element of thickness dx a distance x into the slab will be

$$dI(0) = \varepsilon dx \, e^{-\tau x/X} \tag{4.9.7}$$

so integrating through the slab

$$I(0) = \frac{\varepsilon X}{\tau}(1 - e^{-\tau}) = I_0 \left(\frac{1 - e^{-\tau}}{\tau}\right) \tag{4.9.8}$$

where I_0 would be the intensity in the absence of dust. Thus the face-on absorption is

$$A = -2.5 \log\left(\frac{1 - e^{-\tau}}{\tau}\right) \tag{4.9.9}$$

instead of simply $-2.5 \log e \, \tau$ (equation 2.7.5).

If we tilt the galaxy by an angle i then we simply change the effective depth by a factor $\sec i$ and

$$I(i) = \frac{\varepsilon X}{\tau}[1 - e^{-\tau \sec i}] = I(0)\left[\frac{1 - e^{-\tau \sec i}}{1 - e^{-\tau}}\right] \tag{4.9.10}$$

For the optically thin case $\tau \ll 1$

$$I(i) = I(0) \sec i \tag{4.9.11}$$

and Holmberg's mean surface brightness becomes independent of i; the total optical flux from the galaxy remains the same since the increased intensity is exactly cancelled by the decrease in area of the image.

For the optically thick case, $\tau \gg 1$, on the other hand,

$$I(i) = I(0) \tag{4.9.12}$$

and Holmberg's projected surface brightness and the total optical flux vary as $\cos i$. This occurs because we always see just the top layer of stars, ~ 1 optical depth thick, regardless of inclination. Notice that this means that Holmberg's measure of surface brightness will decrease only by a moderate amount for an arbitrarily optically thick slab, in fact in almost the same way as for a fairly optically thin screen. In the case of the 'sandwich' models, the variation with inclination is even smaller, and we can easily find a physical thickness for an optically thick dust layer which closely mimics the behaviour of Holmberg's optically thin screen or, indeed, an optically thin slab. Thus we cannot make any simple deductions concerning the overall optical thickness or otherwise of typical discs, just from their surface brightnesses. To complicate things further, we should also note that in our calculation we have considered only absorption and not scattering of the photons. Realistic radiation-transfer models show that the latter can have an important effect.

As the light absorbed in the optical should reappear somewhere in the far infra-red, we might hope to use the FIR to optical ratio L_{FIR}/L_B to investigate the dust optical depth. However, this ratio also depends on how clumpy the absorbing dust is, for instance it may mainly surround star forming regions, leaving the rest of the disc relatively clear. For our Galaxy we know that the optical depth through the disc at the solar position must be of order 0.5 (as there is $\sim$0.2 magnitudes of extinction towards the poles) and will almost certainly be higher, $\tau \geq 1$, nearer to the centre of the disc. This may reflect the situation in the majority of spiral discs.

We might expect the optical depth due to dust to scale with the amount of heavy elements from which the dust can form. If typically of order half of the metals in the ISM gas are depleted onto grains, we can estimate an optical depth

$$\tau \sim 4Z\left(\frac{N_H}{10^{20}\,\mathrm{cm}^{-2}}\right) \tag{4.9.13}$$

This is nicely consistent with equation (4.6.16) if $Z \sim 0.01$.

4.10 Spiral structure

The spiral arms correspond to denser than average regions of the interstellar medium. Indeed, most of the densest (and coldest) gas in spiral galaxies lies in the spiral arms. In large spirals this is further concentrated into the GMCs (our Galaxy contains a thousand or more) with masses more than $10^5 M_\odot$, diameters above 10 pc and densities of order 2×10^8 molecules per m^3. Molecular Clouds also exist outside the arms but are much smaller, only a couple of parsecs across and with masses $<100 M_\odot$.

Giant Molecular Clouds tend to have even denser cores containing larger molecules and this is where new generations of stars are formed. Consequently we also find star-forming regions and HII regions strung out along the arms. This accounts for the prominence and blue colours of spiral arms seen in external galaxies. The dense material also contains considerable amounts of dust, so we also see dust lanes along the inner edge of the spiral arms. In our Galaxy we can trace the spiral arm pattern by using the same tracers, though this is now more difficult as we need to be

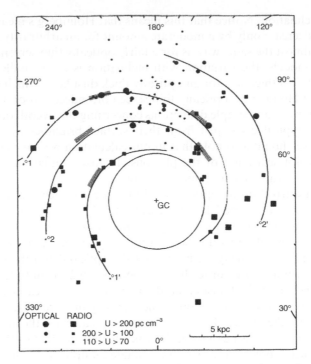

Figure 4.25 The Galaxy's spiral arms as traced by HII regions (Reproduced with permission from Georgelin and Georgelin, 1976). The Sun lies between the arms labelled 1 (Sagittarius-Carina) and 2′ (the Perseus arm).

able to determine distances to the individual GMCs, OB associations, open clusters or HII regions. For the GMCs and HII regions this usually requires measurements of their radial velocities and use of the Galaxy's rotation curve to place them appropriately in 3-D (for example, Figure 4.10).

Nevertheless, starting with the work of Westerhout and Schmidt in 1954, a reasonably detailed picture of the spiral arms of our galaxy has been obtained over the years. The Sun lies near the outer edge of the Sagittarius-Carina Arm (so named because its main star-forming regions lie in the direction of those constellations; the Galactic Centre is also in the direction of Sagittarius, but much further away than the spiral arm). This arm can be traced for almost a full turn around the Galaxy. Slightly further out than the Sun we see a short filamentary arm segment, the Orion Spur, but the next major arm is the Perseus Arm about 2 kpc beyond the Sun.

4.10.1 Density waves

We know that spiral galaxy discs are in differential rotation, that is, the angular velocity Ω is not constant across the galaxy. If the spiral arms were 'material' arms in the sense that the same particles always formed the arms, then the differential rotation would quickly wind them up (the central parts having much higher angular velocity) in just a few rotation periods, that is, a few hundred million years. Some spiral structure, for example, in flocculent spirals, may be short-lived and simply regenerated by the shearing of star-forming regions by the differential rotation itself. However, the fact that the majority of disc galaxies show spiral structure at the present time probably requires some mechanism for maintaining the spiral shape. This also imples that different particles (stars and gas) must be in the arms at different times.

This in turn suggests that the spiral may be due to some sort of wave phenomenon. For ordinary water waves, the wave *crests* move across a lake, but the individual water molecules do not, they merely rise and fall as the wave passes.[8] There are a number of possible mechanisms for generating spiral patterns. The 'density wave theory' posits that the gravitational attraction between the stars at different radii counteracts the tendency of a spiral to wind up and in fact reinforces a pattern which rotates with a fixed pattern speed Ω_p. A star orbiting at radius R passes through such a pattern with frequency $m(\Omega_p - \Omega(R))$, for an m-armed pattern. The reinforcement can only occur if the perturbing frequency (i.e. frequency of arm passages) is less than the epicyclic frequency $\kappa(R)$. A continuous spiral wave can therefore only exist between the 'inner Lindblad resonance' (ILR) and 'outer Lindblad resonance' (OLR), defined as the radii at which $\Omega(R) = \Omega_p + \kappa/m$ and $\Omega_p - \kappa/m$ respectively. Outside the OLR or inside the ILR, the pull of the extra density in the arms occurs more frequently than their epicyclic rate and they are unable to respond and move to reinforce the spiral density enhancement. Recall that for our Galaxy near the Sun $\kappa \simeq 1.4\Omega(R)$ or $\kappa \simeq 36 \, \mathrm{km \, s^{-1} \, kpc^{-1}}$. For a two armed spiral the pattern speed must therefore be between $0.3 \, \Omega$ and $1.7 \, \Omega$.

Except around corotation, where $\Omega(R) = \Omega_p$, the speed at which material enters the arm, $R(\Omega(R) - \Omega_p)$, is supersonic. This drives a shock wave at the front edge of the arm, further increasing the density and helping to instigate the star formation process. Notice that since OB stars have lifetimes of only a few Myr, their position closely tracks that of the shock front and they die out almost as soon as it has passed. Dust lanes appear on the trailing (concave) side, where the gas enters the arm.

Notice that even without the winding up process for matter arms, we might expect the 'heating' of stars in the disc to wash out the spiral pattern. The stars' increased peculiar velocities, as they get older, should allow them to move more easily away from the potential minima in the arms, decreasing their density contrast (even if new stars are still forming with low peculiar velocities close to the spiral itself). This again reinforces the need for some sort of method for continually regenerating the spiral.

In general, energy must be drawn from the disc rotation to maintain the spiral pattern. Many spirals with well-developed arms have nearby companions – a classic case is M51 which has a small companion at the end of one arm – or strong bars. The non-axisymmetric gravitational field may then help to drive the spiral. Simulations suggest that another possibility is the 'swing amplification' mechanism. If Ω_p is large enough so that there is no ILR, waves can travel 'through' the centre. As they do so, they switch from trailing to leading, but then the differential rotation winds them back to a trailing form and the process repeats, amplifying the wave in the process.

---

[8] This is perhaps more obvious with a 'Mexican wave' going round a football stadium.

4.10.2   Bars

The sort of disc heating mentioned above tends to lead also to the formation of a bar. Indeed spiral discs are generically prone to 'bar instability', that is, perturbations tend to end up producing a central bar. According to calculations made by Alar Toomre in 1963, a disc of surface density Σ is unstable if the quantity

$$Q = \frac{\kappa c_s}{\pi G \Sigma} < 1 \qquad (4.10.1)$$

where c_s is the sound speed. Interestingly, this instability of discs leading to bar formation was another step on the path to the current picture of spiral galaxies dominated by dark matter. In 1973, Jerry Ostriker and Jim Peebles demonstrated that pure stellar discs should be so unstable that all discs should have degenerated into a bar configuration, contrary to observation. To prevent this wholesale production of bars, they suggested that the discs must be surrounded by massive dark halos.

Inside the bar, stars and gas effectively orbit *along* the bar, that is they stay within the figure of the bar while the bar rotates about the centre of the galaxy with fixed angular-pattern speed Ω_{bar}. Bars usually end close to the point where Ω_{bar} is equal to the circular velocity $\Omega(R)$, that is, corotation between the bar pattern and the stars' orbits. Orbits outside the bar are then perpendicular to the ends of the bar. Bars are also important in ferrying material into the centres of galaxies. Gas needs to lose angular momentum to reach the centre and the strongly non-central gravitational force due to a bar allows this. Dust lanes form along the leading edges of the rotating bar, opposite to the trailing arms.

4.11   Star formation

As we observe HII regions and bright blue (short-lived) stars in the spiral arms, it is evident that these are important areas for star formation. Since star formation itself takes place in the dense cores of molecular clouds, the earliest stages can best be observed at mm wavelengths. Star forming *regions* show a range of features of different ages, detectable in different ways, though. When the clouds collapse they heat up their dust content from very low temperatures to around 30–50 K, leading to far infra-red emission at 100 μm. As the proto-stars form at the centre, the dust near them reaches temperatures of 1000 K, thus emitting in the near infra-red at 3 μm while the surrounding layers are still at only 100 K. When a star reaches the main sequence (assuming it is massive and therefore hot enough) it will ionise its surroundings, forming a compact HII region detectable at microwave radio wavelengths. As the hot ionised region expands, the dust cools back to FIR emitting temperatures before dissipating and finally leaving the stars visible to view at optical wavelengths as an OB association or open cluster. Notice that once the star formation has taken place, the stars start to feed back energy, and later material, into the ISM, allowing the process to repeat. The feedback occcurs both via the stars' radiation (heating and ionising photons) and mechanically from mass loss as the stars age, or more spectacularly in supernova explosions which can each deposit $\sim 10^{44}$ J into the surrounding medium.

The nearby Orion Nebula region – Orion's sword in the well-known constellation – shows all these features. Orion Molecular Cloud 1 (OMC1) is a dense molecular core of several solar masses behind the main nebula and is the least evolved (youngest) component. There are numerous infra-red sources, including the Kleinman–Low and Becklin–Neugebauer objects in the core of OMC1, the latter apparently being an obscured $10 M_\odot$ B0 star just arriving on the main sequence. Other molecular material exists in parsec scale clumps containing $10–100 M_\odot$. The stars of the Trapezium represent a 10^6-year-old OB associaation and are the brightest of several hundred stars in a 0.3-pc radius region embedded in an HII region which is now eating away at the molecular clouds.

The molecular complexes thus show the signs of 'sequential star formation', as the star formation activity spreads across the region. The formation of only 10 or so OB stars is sufficient to dissociate the H_2 via their UV photons and generate an expanding HII region as we saw earlier. Shock waves propagate ahead into the molecular clouds, though, compressing the gas and instigating gravitational collapse and the formation of further new stars on a time scale of order a million years. The Trapezium is thus the youngest (and smallest) of the OB associations in Orion and adjoins what is left of the molecular gas, which may all be gone in another few tens of millions of years.

4.11.1   The Jeans mass

To get an idea of scales, we can consider the equilibrium, or lack of it, of a gas cloud. If we have a molecular cloud of size r and density ρ at temperature T, in equilibrium, then we can use the standard equation of hydrostatic balance between inward gravitational force and outward pressure force

$$\frac{dP}{dr} = -\frac{GM(r)\rho}{r^2} \tag{4.11.1}$$

At the edge of the cloud $M = 4\pi r^3 \rho/3$ and we can approximate dP/dr by P/r so

$$P \simeq \frac{4\pi}{3} G\rho^2 r^2 \tag{4.11.2}$$

Also, for a perfect gas

$$P = \frac{\rho k_B T}{\mu m_p} = \rho c_s^2 \tag{4.11.3}$$

where μ is the mean molecular weight – close to 2 if the cloud is almost all molecular hydrogen – and c_s is the sound speed. Matching these expressions for P we then obtain

$$r \simeq \left(\frac{3k_B T}{8\pi m_p G\rho}\right)^{1/2} \simeq 10^7 \left(\frac{T}{\rho}\right)^{1/2} \text{ m} \tag{4.11.4}$$

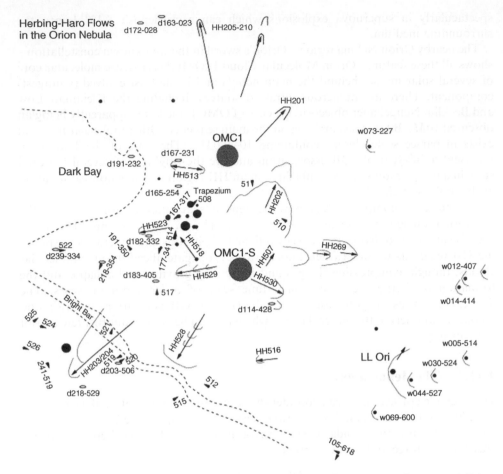

Figure 4.26 A sketch of the components of the Orion star forming complex around Orion Molecular Cloud 1 and the Trapezium (Reproduced with permission from Bally, O'Dell and McCaughrean, 2000). The objects labelled HH are Herbig–Haro objects, regions of excited gas produced by outflows from young stellar objects.

for T in kelvin and ρ in $kg\,m^{-3}$. This value is called the Jeans radius. For a typical GMC, in these units, $T \simeq 10$ and $\rho \simeq 10^{-15}$, so we get $r \simeq 10^{15}\,m$ or about 0.1 pc. The mass contained in such a region is the Jeans mass

$$M_J = \left(\frac{3}{4\pi\rho}\right)^{1/2} \left(\frac{k_B T}{2Gm_p}\right)^{3/2} \simeq 10^{30}\,kg \sim 1M_\odot \qquad (4.11.5)$$

If the same mass is squashed into a smaller space then clearly the gravitational term becomes more important than the pressure term and we might expect the clump to continue to get smaller and denser still. Thus the mass scale on which clumps should be susceptible to collapse seems to be of the right order for forming stars. Notice that it is much less than the mass of a whole GMC. A more sophisticated treatment of the instability, using the equations of gas dynamics and Poisson's equation for the gravitational potential, gives rise to essentially the same Jeans criterion.

4.12   Global star formation

So far we have considered star formation just as a local phenomenon. However, for the overall study of galaxies we often wish to know the total rate at which they are (or have been) forming stars. We can estimate this using the variety of observable effects associated with star formation that we have already met.

4.12.1   Emission lines

Foremost is the Hα emission, which measures essentially the number of ionising photons from young O and B stars. If we assume a standard Salpeter IMF, then we can go from the number of massive stars to the total mass of gas turned into stars over their lifetimes. This therefore measures the current, or at least very recent, star formation rate (SFR), and we will have a relationship of the form

$$\text{SFR} = \frac{L(\text{H}\alpha)}{1.3 \times 10^{34}\,\text{W}}\,M_\odot \text{ per yr} \qquad (4.12.1)$$

where $L(\text{H}\alpha)$ is the luminosity due to photons from the Hα line only (which we can measure from spectra or from narrow-band imaging). For this calculation it has been assumed that stars form with masses in the range between 0.1 and 100$M_\odot$.

Using this relationship we find that typical early type spirals have SFRs from around a tenth to a few $M_\odot$ per yr. Mid-type spirals like our own have SFR $\sim 5 M_\odot$ per yr and this rises to around 10–20$M_\odot$ per yr for Sc-type spirals. The most spectacular 'starburst' galaxies, which are undergoing a sudden galaxy-wide burst of star formation can have SFRs of 100$M_\odot$ per yr or more. Note that for an L_* galaxy and a typical M/L, we might expect a stellar mass a few $\times 10^{10} M_\odot$. This amount of stars can be made in a Hubble time $\sim 10^{10}$ yr by mid- to late-types operating at their current rates. However, this simple argument would suggest that early types must have had higher SFRs in the past (i.e. when they were younger).

There are several observational points that should be borne in mind about these general figures, though. Firstly, they apply to bright galaxies. We might reasonably expect more luminous or larger spirals to have 'more of everything', including more star formation. Thus we may want to normalise the SFR. We can either do this 'after the event' by dividing the derived SFR by, say, the optical area, so that we obtain rates in $M_\odot/\text{yr}/\text{kpc}^2$ (typically of order 0.01), or we can utilise the spectroscopic observations themselves and use the 'equivalent width' of the Hα line. This is defined such that if the total flux in a line is $F(\text{line})$, and the intensity (per Å) in the continuum around the line is $I(\text{cont})$, then the equivalent width (in Å) is

$$EW = \frac{F(\text{line})}{I(\text{cont})} \tag{4.12.2}$$

In other words, a chunk of continuum of width EW would generate the same flux as the line. Spirals typically have Hα EWs from a few up to $\sim$60 Å.

Notice that using EWs is essentially the same as normalising by the (red) luminosity of the galaxy, which in turn is equivalent to comparing the current SFR to the total past SFR (which has built up the population of red stars). Again, vigorously star bursting galaxies will exceed these values for normal spirals and can have Hα $EW > 100$ Å.

Secondly, we have assumed that we measure everything that was emitted. In fact, we know that star formation regions are dusty, so we should allow for the extinction. An average extinction correction in the red of around 1.1 magnitudes (a factor 2.8) is often applied, but we should not be surprised if this value varied quite widely as the star formation and dust are quite clumpy. In principle, we can obtain the extinction in each individual case from what is called the 'Balmer decrement'. If we observe the two Balmer lines Hα and Hβ then their fluxes should be in a specific theoretically calculable ratio. If the observed ratio differs from this, then we can deduce the extra extinction at the wavelength of Hβ compared to that at Hα. Given the wavelengths of the lines this will be roughly $A_B - A_R$, and from this we can easily deduce the total extinction A_R. A more accurate calculation along the same lines provides the extinction at Hα. However, the range of possible extinctions in a single galaxy means that the 'average' extinction may not reflect the real situation. Very dusty regions contribute little to the total light, so are not reflected in the Balmer decrement either, and we may therefore underestimate the actual underlying star formation. Indeed, in order to make the Hα-based SFR agree with those from longer wavelength indicators (below), we need to increase the extinction corrections systematically as the SFR increases. This increases the upper limit for ordinary spirals to around $100 M_\odot$ per yr. If we then integrate over the whole population of star forming galaxies, we obtain a present day star formation density around $0.025 M_\odot/\text{yr}/\text{Mpc}^3$.

Other emission lines can be used as star formation indicators, too. Particularly useful are the oxygen lines [OII]λ3727 and [OIII]λ5007,[9] especially the former, as it is shifted into the optical region of the observed spectra of distant galaxies as the Hα line is redshifted out into the infra-red. For ordinary spirals the equivalent width of the [OII] line is typically one-third of EW(Hα), though again we need to consider dust absorption as this will be much greater at the shorter wavelength.

4.12.2   Other star formation indicators

As a completely different alternative, we can make use of the far infra-red emission. Effectively, we will make a virtue out of a difficulty and choose to measure the emission from the dust which hides the star-forming regions. If we utilise the FIR

---

[9] In this notation the 3727 or 5007 is the wavelength of the line in Ångstroms.

emission integrated from ~10 to 1000 µm, for instance, then this is found to correlate linearly with the SFR, following approximately

$$\text{SFR} = \frac{L(FIR)}{2.2 \times 10^{36}\,\text{W}}\, M_\odot \text{ per yr} \qquad (4.12.3)$$

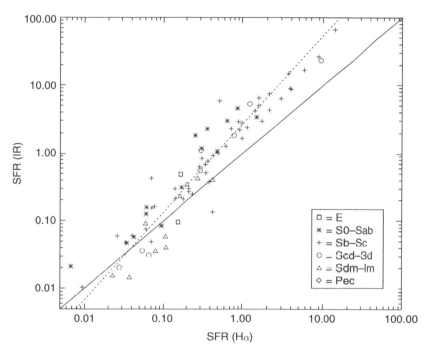

Figure 4.27 A comparison of the SFR deduced from Hα emission and FIR emission (Reproduced with permission from Kewley, Geller, Jansen and Dopita, 2002).

We also know that the HII regions and general ionised gas should be emitting thermal continuum radiation in the radio, and that star forming regions will give rise to supernovae that can accelerate the electrons which produce synchrotron emission. This corresponds quite closely to the situation in the FIR, where there are contributions from dust heated by the general interstellar radiation field – such as the so-called 'infra-red cirrus' in our Galaxy, discovered by IRAS – and from dust in the star-forming regions. Indeed, observationally there is a very tight linear correlation between the FIR flux and the radio flux for all kinds of star-forming galaxies, so we can equally well use, say, the 1.4 GHz flux to determine the SFR. For starburst galaxies, including the ULIRGs, the dust extinction is particularly large (perhaps 2 or 3 magnitudes), and so it is preferable to use the infrared or radio indicators.

In fact, another alternative for ULIRGs and starbursts is to use their X-ray emission. In, say, the 2–10 keV range, the X-rays from ULIRGs are produced primarily by high mass X-ray binaries (HMXBs), while in starbursts there are both HMXBs and LMXBs (i.e. low mass X-ray binaries, the main X-ray emitting population in normal spirals and ellipticals). In both types, the X-ray emission is produced by mass flow from a donor star onto a companion compact object. For the massive ($\geq 8 M_\odot$) donors in HMXBs the lifetime is short, so the HMXBs naturally trace star formation. (The LMXBs, on the other hand, seem to trace the overall stellar mass quite well, with an output $\sim 10^{33}$ W per $10^{10} M_\odot$ of stars.) Normalising to our Galaxy, which has about 50 HMXBs and an SFR $\simeq 3 M_\odot$ per yr, implies a conversion

$$\text{SFR} \simeq \frac{L_X(\text{HMXB})}{10^{32}\ \text{W}} M_\odot \text{ per yr} \tag{4.12.4}$$

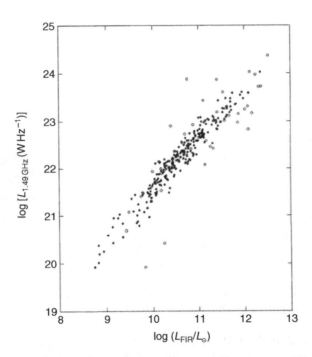

Figure 4.28   The correlation between radio and FIR luminosities (Reproduced with permission from Condon, Anderson and Helou, 1991).

Finally, given that Type II supernovae arise from massive, and therefore short-lived, stars, we might hope to trace star formation via the supernova (SN) rate. For mid-type spirals, this is typically 1–2 SNU, where a 'supernova unit'[10] corresponds to 1 SN per 100 years per $10^{10} L_\odot$ of integrated starlight. For the Galaxy this corresponds to about 1 SN every 30 years (which is also consistent with SNe providing the energy observed in Galactic cosmic rays). This SN rate can be squared with the lack of any SN observed in the 400 years since Kepler's Supernova when it is noted that

---

[10] Not to be confused with the equally eccentric Solar Neutrino Unit which has the same acronym.

interstellar extinction limits us to seeing at most 10% of the Galactic thin disc. Indeed, many more supernova remnants (SNR) – the expanding shells of expelled material – can be observed in the radio. A particular bright example is the source Cas A, an SNR about 3 kpc away and only 250 years old, judged by its size and the expansion speed of the shell. More locally, the Gum Nebula is an old SNR, centred about 700 pc away, but with a shell now ~600 pc in radius.

4.12.3   Densities and timescales

It would seem reasonable to expect that the amount of star formation going on in a galaxy should depend on the amount of available fuel, that is the interstellar gas. The traditional quantification of this idea is the 'Schmidt law'

$$\text{SFR} \propto \Sigma_{\text{gas}}^{n} \tag{4.12.5}$$

where Σ_{gas} is the surface mass density of gas (essentially equivalent to the column density N_H). It was originally proposed that n should be around 2, reflecting the collision rate between pairs of interstellar clouds, but other considerations might lead us to expect that $n \simeq 1$, viz. more gas allows more stars to form. Observations appear to suggest something intermediate, say $n \simeq 1.4$, and the star formation per unit area then follows something like the 'Kennicutt law'

$$\frac{\Sigma_{\text{SFR}}}{M_{\odot}/\text{yr}/\text{kpc}^2} = 2.5 \times 10^{-4} \left(\frac{\Sigma_{\text{gas}}}{M_{\odot}/\text{pc}^2} \right)^{1.4} \tag{4.12.6}$$

It appears that the best correlation is with the total gas density (HI and H_2), as used here, rather than just H_2, even though it is the molecular gas which is the scene of the star formation itself. Note, too that the SFR versus gas density relation also provides a timescale – sometimes called the Roberts time – on which all the existing gas will be used up by star formation at the current rate. This is frequently much less than a Hubble time.

Even starbursts seem to follow the above relation, their very high SFRs being accommodated by surprisingly high gas surface densities $\sim 10^2$ to $10^5 M_{\odot}\,\text{pc}^{-2}$ compared to the 1 to $100 M_{\odot}\,\text{pc}^{-2}$ for normal spiral discs.

The power, close to 3/2, can be understood if the SFR depends not only on the gas density (i.e. ρ_{gas}) but also (inversely) on the time scale for forming stars. This should be of order of the 'free fall time', t_{ff}, which is the time for a perturbation to collapse. We can easily determine this by imagining a particle falling from the edge of a perturbation of size r to the centre along an infinitely thin 'ellipse', which must have semi-major axis $a = r/2$. But we know from Kepler's third law of planetary motion (as explained by Newton's theory of gravity), that a is related to the period of the orbit, P, by

$$P^2 = \frac{4\pi^2}{GM} a^3 \tag{4.12.7}$$

and of course $M = 4\pi\rho r^3/3$ where ρ is the initial density: so given that the time to fall to the centre must be $P/2$, we have finally

$$t_{ff} = \left(\frac{3\pi}{32G\rho}\right)^{1/2} \tag{4.12.8}$$

(which is also of order the Jeans radius over the sound speed). Thus

$$t_{ff} \propto 1/(G\rho)^{1/2} \tag{4.12.9}$$

(as indeed it must be, just on dimensional grounds) and we can expect

$$\text{SFR} \propto \frac{\rho_{gas}}{t_{ff}} \propto \rho_{gas}^{3/2} \tag{4.12.10}$$

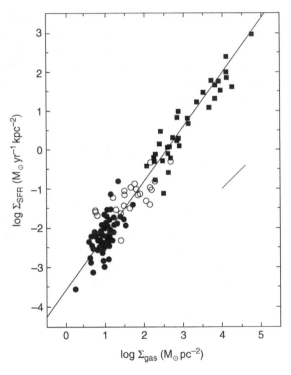

Figure 4.29 The star formation rate per unit area as a function of gas surface density (Reproduced with permission from Kennicutt, 1998).

Alternatively, the relevant star formation timescale might be a dynamical time associated with the rotation period of the disc, for instance the time between arm passages, if star formation is induced by the compression wave as gas enters the arm. We might then have

$$\text{SFR} \propto \Omega \rho_{\text{gas}} \tag{4.12.11}$$

where Ω is the angular velocity of the gas disc.

We can explore the star formation in more detail *within* a given galaxy. Two features are apparent if we do this. First, there appears to be a minimum column density below which star formation does not take place, at around $4M_\odot\,\text{pc}^{-2}$, and secondly there is a Schmidt-type relation above that, roughly

$$\Sigma_{\text{SFR}}(R) \propto \Sigma_{\text{gas}}(R)^{1.3} \tag{4.12.12}$$

The threshold value can be understood if we recall Toomre's criterion for stability $Q < 1$ (equation 4.10.1), which corresponds to a critical density

$$\Sigma = \frac{\kappa c_s}{\pi G} \tag{4.12.13}$$

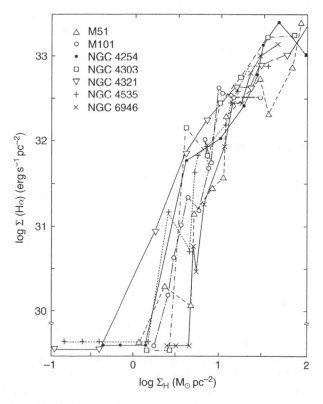

Figure 4.30   The radial variation of star formation compared to the gas density, for a sample of spiral galaxies, demonstrating the threshold below which star formation appears to be cut-off (Reproduced with permission from Kennicutt, 1989).

Observationally it does appear that star formation (or at least the presence of HII regions) is limited to regions of galaxies where the density is above this threshold. Also, if we set the dynamical time at any radius as simply $2\pi R/V(R) = 2\pi/\Omega$, then the observed radial variation in SFR follows Σ_{gas}/t_{dyn} remarkably well, with

$$\Sigma_{SFR} \simeq 0.017\,\Sigma_{gas}\Omega \qquad (4.12.14)$$

This suggests that about 10% of the gas is turned to stars per orbital period.

Locally, the average star formation (over a 10^8–10^9 yr timescale) near the Sun corresponds to somewhat less than 1 solar mass star per square parsec per Gyr. Given the disc scale height this will be about 2 stars per 10^3 cubic parsecs per Gyr. Even modest-sized galaxies can contain some regions with high SFR, though. For instance, the central 10 pc core of the 30 Dor region of the Large Magellanic Cloud houses some $10^4 M_\odot$ of young stars and has a star formation rate (over a very small area, of course) equivalent to $\sim 100 M_\odot/\text{yr}/\text{kpc}^2$.

The environment a galaxy finds itself in can have a strong influence on its star formation. We can note two particular instances of this. On the one hand, strongly interacting galaxies, exemplified by ULIRGs, have very high SFRs. On the other hand, spiral galaxies in clusters like Virgo appear to have lower SFRs – and lower gas masses – than galaxies of the equivalent type in the field. These 'HI deficient' galaxies may indicate that gas consumption has been speeded up in the cluster environment, possibly as a result of effects induced when spirals 'fall' in to the cluster from the periphery. In general, the large redshift surveys have shown that the SFR decreases with increasing density of the local environment in which a galaxy finds itself.

4.13   Chemical evolution

Closely linked with the star formation is the chemical evolution of galaxies, that is the production of heavy elements in stars. Nucleosynthesis in the early universe ('Big Bang nucleosynthesis') would have produced around 25% by mass in helium and a smattering of deuterium and other light elements, but all the other elements owe their existence to nuclear processing inside stars. In order to model the gradual enrichment of the ISM in a star-forming galaxy, we need to make some simplifying assumptions.

4.13.1   Closed box models

In particular, we will consider what are called 'closed box models'. This means that no interstellar gas flows into or out of our system. Furthermore, we will assume that all the stellar mass returned to the ISM (for instance following supernova explosions) comes from short-lived stars, so that we can use the 'instantaneous recycling approximation', that is relative to the overall evolutionary timescale this mass is processed instantaneously, spending a negligible time actually in a star. We will also assume that the mass lost from the stars is perfectly mixed into the rest of the ISM, so that its chemical composition is uniform.

Consider then a unit mass of a galaxy and let $g(t)$ be the mass (i.e. fraction) which is in gas at any time t. Counting time from the epoch when the galaxy formed from a gas cloud, we thus have the boundary condition $g(0) = 1$. Similarly let $Z(t)$ be the metallicity in the gas at time t. Also set $S(t)$ to be the total mass which has been 'astrated', that is has gone into stars which have formed up to time t (including the mass recycled back to the ISM). The fraction of this mass which remains in long-lived stars or in remnants like white dwarfs or neutron stars is denoted by α, so $1 - \alpha$ is obviously the fraction ejected by the stars, sometimes called the 'return fraction'. For a normal IMF, stellar evolution theory tells us that α should be around 0.8. Finally we need the 'yield'. If we let the mass of heavy elements produced *and* returned to the ISM be p', then $p = p'/\alpha$ is the yield, the mass of heavy elements formed per unit mass still locked up in stars. Theoretical models suggest $p \simeq 0.01$, but its value depends not only on the IMF and stellar evolutionary processes but also on the mechanics of supernova explosions, so may be better determined empirically. In fact, different elements are produced in different ways in different sorts of stars, so in principle we should consider the yield of each element separately, but we will not worry about that complication here. In particular, we will only consider elements – called 'primary elements' – whose production does *not* depend on the existing metal content, and so we can take p to be constant.

With these definitions, the rate of production of metals must follow

$$\frac{\mathrm{d}(gZ)}{\mathrm{d}S} = p' - \alpha Z \tag{4.13.1}$$

where the first term on the right is the amount of new metals created and ejected into the ISM while the second term represents the existing metals which have now been lost into new long-lived stars. But we also have

$$g = 1 - \alpha S \tag{4.13.2}$$

so

$$\frac{\mathrm{d}(gZ)}{\mathrm{d}g} = -p + Z$$

or

$$g\frac{\mathrm{d}Z}{\mathrm{d}g} = -p \tag{4.13.3}$$

This then has the simple solution

$$Z(t) = p \ln(1/g(t)) \tag{4.13.4}$$

Thus we expect to see the metallicity of the gas in star-forming galaxies increase logarithmically as the gas fraction goes down.

Now consider the stars. Since we have assumed instantaneous perfect mixing of stellar ejecta back into the ISM, the stars forming at any particular time form with metallicity $Z(t)$ and, equivalently, all the stars which formed before time t must have had metallicity less than $Z(t)$, the contemporary gas metallicity. Thus the mass of stars with metallicity less than Z_* is

$$s(<Z_*) = 1 - g(t) = 1 - e^{-Z_*/p} \qquad (4.13.5)$$

(Note the distinction here that s is the mass *in* stars at a given time, whereas S is the mass that *has been* in stars). Differentiating this equation, we expect the number of stars with different metallicities to follow

$$n_s(Z_*) \propto e^{-Z_*/p} \qquad (4.13.6)$$

We can also determine the (mass weighted) mean stellar metallicity at time t as

$$<Z_*(t)> = \frac{1}{s(t)} \int Z \mathrm{d}s = \frac{1}{1-g(t)} \int \frac{Z}{p} e^{-Z/p} \mathrm{d}Z \qquad (4.13.7)$$

where the integrals are over all times up to t and we have used

$$\mathrm{d}s = -\mathrm{d}g = g\mathrm{d}Z/p = e^{-Z/p}\mathrm{d}Z/p \qquad (4.13.8)$$

Integrating this by parts and taking the limits 0 and $Z(t) = p \ln(1/g(t))$ we have

$$<Z_*(t)> = \frac{1 - g(t) - g(t) \ln(1/g(t))}{1 - g(t)} p \qquad (4.13.9)$$

Notice that as star formation goes to completion, i.e. as the gas content approaches zero, the mean stellar metallicity tends to the yield p.

One of the clearest predictions of the simple closed box model is that a substantial fraction of stars should have very low metallicity. If we look at the stars in the bulge of our Galaxy, they appear to follow the above exponential distribution quite well if we take $p \simeq 0.7 Z_\odot$, a reasonable value according to theoretical models of stellar nucleosynthesis. The bulge may thus have managed to turn all its initial gas into stars.

In the disc of the Galaxy, on the other hand, there appear to be very few, old (i.e. long-lived), low metallicity stars, say with $Z < Z_\odot/10$. The present gas fraction near the Sun is about 1/4 and the average metallicity of the gas is probably around $0.7 Z_\odot$. Thus $Z(t_0) \simeq 0.7 Z_\odot \simeq p \ln 4$ which implies $p = 0.5 Z_\odot$, again a perfectly reasonable value theoretically. (Notice that the Sun is more metal rich than the gas, even now: it was therefore very much more metal rich than the average for the gas 4.5 Gyr ago when it formed.[11])

---

[11] It has been hypothesised that this richness in heavy elements in the proto-solar cloud may have been instrumental in the formation of planets.

If we look at the Z distribution, we should have

$$s(<Z_\odot/10) = 1 - \exp(-Z_\odot/10p) = 1 - e^{-1/5} \simeq 0.2 \qquad (4.13.10)$$

so 20% of local old-disc stars should have $z < Z_\odot/10$, whereas virtually none are observed. This problem was first noted for (suitably long-lived) G stars like the Sun, so is known as the 'G-dwarf problem'.

There are several possible solutions. Firstly, the initial gas out of which the disc formed may not have had primordial composition. Perhaps a generation of Population III stars had already polluted the gas with metals, up to some initial value Z_0 (the so-called 'prompt initial enrichment' or PIE model). This would change the expected distribution to

$$s(<Z_*) \propto \exp(-(Z_* - Z_0)/p) \qquad (4.13.11)$$

for $Z_* > Z_0$. To solve the G-dwarf problem this way requires $Z_0 \simeq 0.15Z_\odot$ (but a lower yield, $p \simeq 0.4Z_\odot$). This obviously removes all the very low Z_* stars and predicts that about 20% of all stars should be below $Z_\odot/4$, in reasonable agreement with observation.

4.13.2   Gas flows

However, a more likely explanation is that one of our model assumptions breaks down. The instantaneous return of heavy elements to the ISM is certainly questionable and more detailed models allow for the lifetimes of the stars which eventually become supernovae, for instance. In particular, not all heavy elements will be produced at the same rate. Oxygen is produced in Type II supernovae (SNII) which have progenitor masses above $8M_\odot$ and lifetimes of only 100 Myr, but iron is produced by SNIa, which result from mass flow on to white dwarfs in binaries on

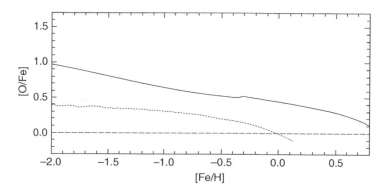

Figure 4.31   The expected variation of the element ratio [O/Fe] with [Fe/H] for bulge (continuous line) and disc (dashed line) stars. (Models from Matteucci, Romano and Molaro, 1999).

time scales ~1 Gyr. Also, we have ignored mass lost by the longer lived, lower mass stars, which turns out to be an important source of carbon and nitrogen. Evidence for such delayed production of some elements can be seen in, for example, the [O/Fe] ratio in stars of different overall metallicities; old halo stars, in particular, are over-abundant in O (and Mg) compared to Fe. In addition, we have assumed perfect (and instant) mixing of the mass lost into the existing ISM. We can guess from the wide spread of observed metallicities amongst stars of similar ages that this may be a poor approximation.

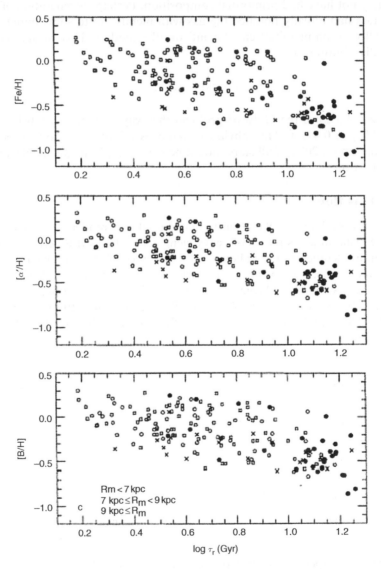

Figure 4.32 The age–metallicity relation for Galactic stars (Reproduced with permission from Edvardsson *et al.*, 1993). α' in the middle panel represents the average of Ca and Si.

However, the most critical assumption is the initial one of a closed box. If we allow for inflow or outflow of gas into or out of the 'box' (i.e. the Galactic disc), then the equations of chemical evolution contain extra terms and allow for a much wider range of metallicity distributions. For instance, for an inflow model, the differential equation for the gas mass g (per unit mass of *initial* gas) becomes

$$g\frac{\mathrm{d}g}{\mathrm{d}S} = -\alpha(g - f) \tag{4.13.12}$$

where f quantifies the accretion rate ($\mathrm{d}A$) compared to the star formation rate ($\mathrm{d}S$), via $\mathrm{d}A/\mathrm{d}S = \alpha f/g$. For constant f we can then show that the simple law

$$\frac{Z}{p} = \ln\left(\frac{1}{g}\right) \tag{4.13.13}$$

becomes

$$\frac{Z}{p} = \frac{g-f}{g}\ln\left(\frac{1-f}{g-f}\right) + \frac{f(1-g)}{g(1-f)} \tag{4.13.14}$$

Physically, the low initial mass of gas (compared to the final mass) means that only a few stars form at low Z, and even if the inflowing gas is relatively unenriched, the net effect is to create a larger fraction of stars from higher metallicity gas.

Notice that none of the above depends directly on the SFR or its history, but merely on the end result in terms of the remaining gas fraction. We can determine how metallicity should increase with time if we fold in an assumed star formation law, such as the Schmidt law, which we can write in the present context as

$$\frac{\mathrm{d}S}{\mathrm{d}t} = k g^n \tag{4.13.15}$$

This implies, for the closed box model (from equation 4.13.1),

$$-\frac{1}{g}\frac{\mathrm{d}g}{\mathrm{d}t} = \frac{1}{g}\alpha\frac{\mathrm{d}S}{\mathrm{d}t} = \alpha k g^{n-1} \equiv \frac{1}{\tau} \tag{4.13.16}$$

where τ is a characteristic timescale for the star formation. Thus (from 4.13.7)

$$\frac{\mathrm{d}Z}{\mathrm{d}t} = \frac{p}{\tau} \tag{4.13.17}$$

For $n = 1$, τ is a constant so $Z = pt/\tau$ and $g = e^{-t/\tau}$.

In our Galaxy, we can directly determine ages (and hence birth dates) and metallicities of disc stars from detailed observations of their individual colours and spectral line strengths. Although with the large scatter at any deduced age which was mentioned above, there is clearly a trend in the expected sense of increasing metallicity for younger stars, the 'age–metallicity relation' or AMR. The oldest disc stars

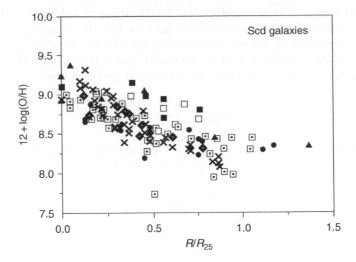

Figure 4.33 Variation of metallicity with radius in some nearby galaxies (Reproduced with permission from Vila-Costas and Edmunds, 1992).

(plotted at ages around 15 Gyr in the figure)[12] have metallicities of $0.2-0.3Z_\odot$ while recently formed stars (say less than 2-Gyr-old) have $0.7 \leq Z \leq 2Z_\odot$.

4.13.3   Radial gradients

The metallicity of disc stars also varies radially. Stars closer to the centre are typically more metal rich. This sort of metallicity gradient is also seen in other spirals and can be explained in terms of the dependence of SFR on gas density Σ_{gas} (which also varies radially). For instance, if we write the SFR as

$$-\frac{1}{\alpha}\frac{d\Sigma_{gas}}{dt} = k\,\Sigma_{gas}^n \tag{4.13.18}$$

then if Σ_{tot} is the total surface density of stars plus gas, we have

$$\frac{\Sigma_{gas}}{\Sigma_{tot}} = -\alpha kt \quad \text{if } n = 1 \tag{4.13.19}$$

and

$$\frac{\Sigma_{gas}}{\Sigma_{tot}} = [1 + (n-1)\alpha kt\Sigma_{tot}^{n-1}]^{-1/(n-1)} \quad \text{if } n \neq 1 \tag{4.13.20}$$

Our earlier equations for metallicity in terms of the gas fraction then imply that if $n = 1$

---

[12] The figure uses older models of stellar structure which have ages larger than the currently accepted age of the universe, but the *relative* ages should still be correct.

$$Z = \alpha kpt \tag{4.13.21}$$

independent of Σ (i.e. no gradient). But if $n > 1$ we obtain

$$Z = \frac{p}{n-1}\ln[1 + (n-1)\alpha kt\Sigma_{\text{tot}}^{n-1}] \tag{4.13.22}$$

which we could also write as a (more complicated) function of the stellar surface density, corresponding to the observed surface brightness.

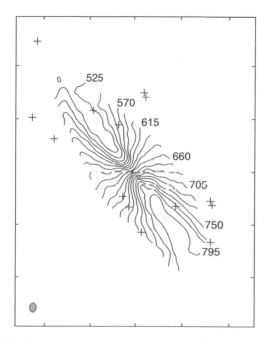

Figure 4.34 The 'spider diagram' for NGC3198 (Reproduced with permission from Begeman, 1989). The contours are lines of constant gas velocity, as seen by the observer.

4.14   Rotation of the gas

We should next consider the overall dynamics of the interstellar gas. As the gas rotates along with the stellar disc, and it is trivial to measure the redshift of the radio 21-cm line, we can easily measure the radial velocity at different points across a galaxy and determine rotation curves of external spiral galaxies in this way. Indeed, for a well-resolved galaxy, if we plot lines of constant radial velocity for an inclined rotating disc, we obtain what is called a 'spider diagram', with roughly parabolic contour lines on either side of the minor axis (where we see no radial velocity apart from the overall recession velocity of the galaxy).

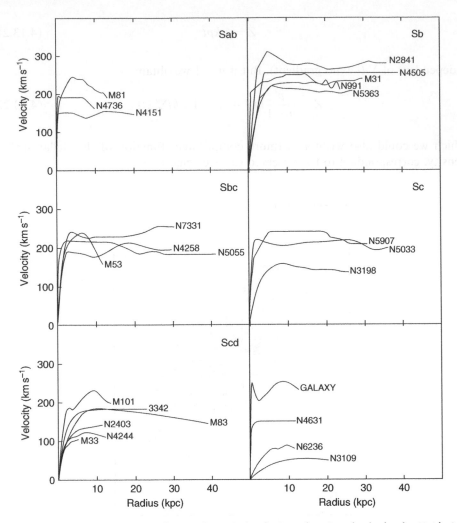

Figure 4.35 HI rotation curves for nearby spiral galaxies, showing the lack of a Keplerian fall-off at large radii (Reproduced with permission from Bosma, 1986).

Although the spatial resolution of the observations may not be as good as for those in the optical (because of the usual diffraction limit $\sim\lambda/D$ for radiation of wavelength λ observed with a telescope of diameter D), the HI is frequently detectable far from the galaxy centres. Deep HI observations can reach column density levels around 10^{19} atoms per cm^2 or 0.1 $M_\odot$ pc^{-2} which occur typically at 2 or 3 optical radii.

Observations like these, assembled in the late 1970s, showed that even beyond the visible stellar disc the rotational velocity *still* fails to drop in Keplerian fashion. Indeed, the rotation curves generally remain flat out to the limits of the observations. From the simple $V^2 \propto M/R$, this obviously requires that the mass inside radius R increases linearly with R even where there are few (or no) stars. As the gas itself provides only a small contribution to the required mass, this was the third major indicator of the presence of missing mass, and provided the main impetus behind serious theoretical consideration of dark matter in the 1980s.

If we make reasonable estimates for the M/L ratio for the stars in the bulge and disc of a spiral, and allow for the ISM mass, then whatever shortfall there is compared to what is needed to generate the observed $V(R)$ must be due to the dark matter. Even with a 'maximum disc model', that is, choosing the largest M/L for the disc that does not overpredict the rotation curve near the centre, it is found that generally a 'dark halo' with a density distribution not far from that of an isothermal sphere is required to match the rotation curve. (A singular isothermal sphere (SIS) has the required $\rho \propto 1/r^2$ to generate a mass increasing in proportion to radius, but has the more unpleasant characteristics of an infinite density at the centre and divergent total mass.)

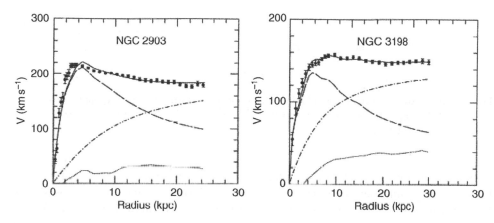

Figure 4.36   Fits of galaxy rotation curves by three mass components: stars – dashed line, gas – dotted line and halo – dash-dot line (Reproduced with permission from Begeman, Broeils and Sanders, 1991).

For instance, a spherical halo density distribution with

$$\rho_h(r) = \frac{V_h^2}{4\pi G} \frac{1}{r^2 + a_h^2}$$

(4.14.1)

gives a rotation curve

$$V^2(R) = \frac{GM_h(R)}{R} = V_h^2 \left(1 - \frac{a_h}{R} \tan^{-1} \frac{R}{a_h}\right)$$

(4.14.2)

which asymptotes to $V(R) = V_h$ at large R. At small R, V rises nearly linearly with R, as also observed. For most spirals, 50 to 90% of the mass out to large radii (tens of kpc) must be dark and the overall M/L ratios rise to 10 or 20 as we move outwards. The mass of our Galaxy follows

$$M(<R) \simeq 10^{11} (R/10\,\mathrm{kpc}) M_\odot$$

(4.14.3)

from around 3 to at least 20 kpc.

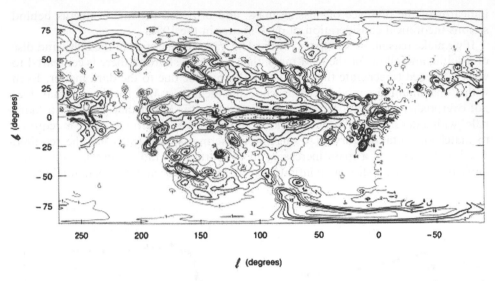

Figure 4.37 An HI survey map of Galactic emission in a particular velocity range (from Stark *et al.*, 1992). The 'cut-out' regions towards the edges of the map are too far south to be observed in this survey.

In fact, computer 'N-body' simulations of the mutual gravitational attractions between large numbers of dark matter 'particles' suggest that dark halos should have a common form, called the NFW profile (after Navarro, Frenk and White). This has

$$\rho_{\text{NFW}} \propto \frac{1}{(r/r_s)(1 + r/r_s)^2} \qquad (4.14.4)$$

where r_s is a core scale size. Note that this has a flatter slope than the SIS profile near the centre and a steeper one at large r.

4.14.1 Rotation of gas in the Galaxy

For our Galaxy, observing from the inside complicates the determination of the HI rotation curve since we have no direct way of deciding how far away (and hence at what Galacto-centric distance) any particular 21-cm emitting gas actually lies. However, if we assume that the HI is continuously distributed throughout the disc, we can use what is called the 'tangent point method', at least for the inner Galaxy.

Consider looking along some Galactic longitude ℓ. Material close to the Sun will have its velocity mainly across the sky, and so will distant material on the far side of the Galaxy. However, at the point where ℓ is tangent to the circular motion, we will see entirely radial motion and hence the highest radial velocity (either towards or away from us), as can be seen from Figure 4.10. At the tangent point

$$R = R_0 \sin \ell \qquad (4.14.5)$$

and in general

$$\nu_r = V_0 \sin \ell - V(R) \tag{4.14.6}$$

hence

$$V(R_0 \sin \ell) = \nu_r^{max}(\ell) + V_0 \sin \ell \tag{4.14.7}$$

If we want to map the HI in the Galaxy, we can choose to do so for particular velocity ranges (e.g. Figure 4.37). This can be related to the spatial distribution if we know the detailed behaviour of the rotation curve, since we can then work out the distance from us, in direction ℓ, at which ν_r takes a particular value.

Another way of representing the data, peculiar to radio astronomers, is the (ℓ, ν) diagram. Here we plot directly (e.g. as a grey scale or contour map) the amount of HI emission in direction ℓ with radial velocity ν. (Notice that this is a 2-D slice through the Galaxy, usually at a fixed Galactic latitude, often 0°.) Again, we can interpret this in terms of the rotation curve; for instance, the maximum ν at any ℓ corresponds to the tangent point, as before.

If we look specifically at ℓ near 90° (or 270°), so that the tangent point is actually quite close to the Sun, then we can use the same approximations as in section 4.3. The distance $r = R - R_0 = R_0 \sin \ell - R_0$ so

$$\nu_r^{max} = V_0 - \left(\frac{dV}{dR}\right)_{R_0} R_0(1 - \sin \ell) = 2AR_0(1 - \sin \ell) \tag{4.14.8}$$

where A is again the Oort's constant. Thus, measuring ν_r^{max} at given ℓ gives us the value of AR_0 and hence the distance to the Galactic Centre if A is known.

We can, of course, obtain the distance to the centre in other ways, such as Shapley's original method of determining the centroid of the 3-D distribution of tracers such as globular clusters which can be assumed to be symmetrically placed about the centre, or using stellar candles such as RR Lyrae variables. As noted earlier, the standard value usually assumed is 8.5 kpc, though a number of more recent estimates suggest that it may be nearer 8 kpc.

4.15 The Tully–Fisher relation

For distant spiral galaxies we may not be able to spatially resolve the radial velocity curve, but nonetheless we can determine their overall rotation velocities. Unlike (most) optical measurements, radio observations tell us directly the amount of radiation received by the telescope at any particular wavelength or frequency, since the radiation is detected in separate 'channels' of the radio receiver, in exactly the

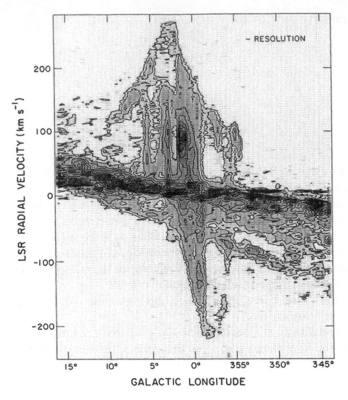

Figure 4.38 An (ℓ, ν) diagram for molecular gas towards the Galactic Centre (from Dame *et al.*, 1987).

same way that you tune an ordinary radio to a specific radio station. For a spiral galaxy which is contained entirely within the radio telescope 'beam' (i.e. it is smaller than the telescope can resolve), we can measure the 'velocity width', that is, the range of velocities (i.e. frequencies) at which we detect 21-cm emission. For a rotating galaxy, this width $\Delta\nu$ is essentially twice the (maximum) rotation velocity, modulated by the inclination angle, i.e. $\Delta\nu \simeq 2V_{\mathrm{max}} \sin i$ and the radio spectrum shows a characteristic 'twin-horned' profile from the approaching and receding material.

The compilation of such rotation velocity measurements led to the discovery by Brent Tully and Richard Fisher in the mid-1970s of a relationship between the linewidth and luminosity following approximately

$$L \propto \Delta\nu^4 \tag{4.15.1}$$

The Tully–Fisher relation is the analogue for spirals of the Faber–Jackson relation for ellipticals and in its original version has essentially the same form. Again, we can understand this directly from the dynamics if we simply assume that $\Delta\nu$ represents the rotation velocity in the outer parts of the galaxy (of radius R), so $\Delta\nu^2 \simeq GM/R$, and that spirals all have roughly the same mass-to-light ratio M/L and surface brightness I, so that $M \propto L \propto IR^2$. Putting all this together implies

$$\Delta v^4 \propto \frac{M^2}{R^2} \propto M\frac{M}{R^2} \propto \left(\frac{M}{L}\right)^2 I L \qquad (4.15.2)$$

Of course, one might rightly object that Δv is really measuring the mass of the dominant dark matter halo and L is related only to the visible stellar component, but nevertheless the relation still holds!

More recent studies, including those undertaken in the infra-red to minimise the effects of dust obscuration, have suggested slightly different exponents. For instance, in the H band

$$L_H \simeq 3 \times 10^{10} L_{H\odot} \left(\frac{V_{\text{max}}}{200\,\text{km s}^{-1}}\right)^{3.8} \qquad (4.15.3)$$

while in the optical B band the exponent is perhaps nearer to 3. Surprisingly, given our simple justification above, lower surface brightness spirals appear to follow the same relation as higher surface brightness objects, presumably requiring some trade-off between M/L and I. V_{max} is usually in the range 100 to $300\,\text{km s}^{-1}$, though a few very massive spirals reach $400\,\text{km s}^{-1}$.

The T F relation remains a vital part of the galaxy astronomer's armoury. Since, like the F–J relation, it relates a distance-independent velocity measurement to the absolute luminosity, we can perform the usual trick of observing the apparent luminosity and using the inverse square law to deduce distances for large numbers of spiral galaxies. It has therefore been one of the major methods in distance scale

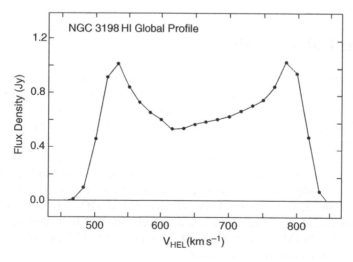

Figure 4.39 The 'twin-horned' velocity profile of the HI emission from NGC3198 (Reproduced with permission from Begeman, 1989).

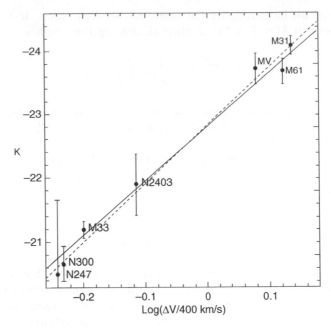

Figure 4.40   The *K*-band Tully–Fisher relation for nearby galaxies (Reproduced with permission from Malhotra, Spergel, Rhoads and Li, 1996).

and Hubble constant work. In addition, it represents one of the key observables which any successful theory of spiral galaxy formation must be able to match, since, as we saw above, it actually provides a link between the dominant dark halo mass and the stars which have formed within it.

4.16   The Galactic halo

We have seen that the rotation curve of the gas in the Galaxy implies that its mass inside 20 kpc is around $2 \times 10^{11} M_\odot$. Even further out than this, we can observe the orbits of satellite galaxies; their velocities suggest a total mass around $10^{12} M_\odot$ out to 100 kpc. This in turn requires $M/L \simeq 50$, much larger than could be accommodated by stars alone.

Although much of the missing mass must be in 'exotic' non-baryonic form (for reasons to be discussed in Chapter 8), it is possible that a significant fraction of it is supplied by invisible but macroscopic objects. These are generically known as Massive Astrophysical Compact Halo Objects (MACHOs), to contrast with the particle physics dark matter candidates collectively referred to as Weakly Interacting Massive Particles (WIMPs). The MACHOs could be small black holes, planet-sized bodies or dim stars such as brown dwarfs.

Although emitting little or no light, these might be detectable by their gravitational effects. According to Einstein's General Theory of Relativity, light passing close to a large mass like a star (or MACHO) will be deflected. This can be seen directly for starlight passing close to the limb of the Sun at a total eclipse, one of the classic tests of General Relativity, first carried out by Arthur Eddington in 1919. The bending of the light ray means that the mass acts like a lens, also amplifying the light of the background source, in the manner of a magnifying glass.

If we consider a large number of background stars – say in one of the Magellanic Clouds – then every now and again a MACHO in our Galaxy's halo should move across the line of sight to one of the stars. When this occurs it will briefly cause the background star to appear to brighten. In the 1990s several teams, including the eponymous MACHO Collaboration, using the refurbished 130-year-old former Great Melbourne Reflector[13] at Mount Stromlo, monitored millions of stars looking for such random brightenings. They have found a number of convincing events which can be differentiated from variable stars because they vary by the same amount at all wavelengths, that is, achromatically. The duration of a lensing event reflects the lensing mass and the results suggest that anything up to 20% of the halo mass may be supplied by dim objects of mass around $0.5 M_\odot$, presumably white dwarfs. However, other considerations, such as star counts, proper motions and the amount of heavy elements that would be produced by the white dwarf progenitors, probably limits the actual contribution to only a couple of per cent.

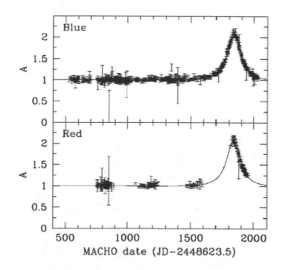

Figure 4.41 Light curves (in blue and red) for the lensing of a star in the SMC by a MACHO (from Alcock *et al.* (the MACHO collaboration), 1997).

---

[13] Unfortunately this was destroyed in the devastating bush fires which struck the Canberra area in 2003.

4.17   The Galactic Centre

Moving in the other way, we have seen that the Galaxy contains a bulge (probably of bar-like shape) about 2 kpc in radius and with a mass around $10^{10} M_\odot$. On a much smaller scale, the Galaxy also possesses a nuclear star cluster of mass around $10^7 M_\odot$ and size ~ 3 pc. The nuclear region contains not only old metal-rich stars, but also very young blue and red supergiants. The central parsec alone appears to contain about 25 blue supergiants with masses between 30 and $100 M_\odot$. The gas from which these must have formed is seen as molecular clouds – the centre contains around $10^6 M_\odot$ of molecular material, probably in a toroidal-shaped region – and HII regions.

Of course, the centre region is obscured in the optical, so we must use infra-red or radio observations. Radio emission at centimetre wavelengths reveals two main regions, labelled Sagittarius (Sgr) A and B (recall that the Galactic Centre region lies in the direction of the constellation Sagittarius). Higher resolution maps show that there are a whole string of radio sources aligned along the Plane in a region about 250 pc across. Their thermal radio spectra reveal these to be HII regions. In the mid infra-red, at 10 µm, we see emission from dust, possibly from regions around forming O stars. Keeping all the hydrogen near the centre ionised requires an input around 2×10^{33} W. The highest resolution X-ray observations from the Chandra satellite give a similar view of the central regions.

Sgr A is very close to the Galactic Centre itself. At high resolution it breaks up into at least three components, East, West and A*, a few parsecs apart. Sgr A East is a non-thermal source, probably a supernova remnant, while Sgr A West appears to be an HII region, also visible as a dusty infra-red source. Sgr A* lies within or behind Sgr A West and is a non-thermal point source (at most a few AU in diameter) which almost certainly marks the actual Galactic Centre. Ionized gas in the very central

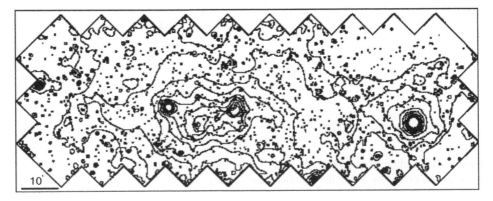

Figure 4.42   A mosaic of Chandra X-ray images of the Galactic Centre region (Reproduced with permission from Wang, 2002).

regions is rotating at several hundred km s^{-1}, indicating a mass of a few $\times 10^6 M_\odot$ inside 0.2 pc.

Dynamical information from even further in towards the centre has recently been obtained in the near infra-red using adaptive optics (to compensate for the turbulence in the atmosphere that blurs the images) on the European Southern Observatory's Very Large Telescope (VLT). Observations with a resolution of just 40 milli-arc seconds (mas) allow us to probe down to scales of 400 AU and it is found that the stellar density keeps on rising right in to this point, approximately as $R^{-1.4}$. Given that the observed stellar radial velocity dispersion also rises, up to a value of about 1000 km s^{-1} at 1$''$ from the centre, this clearly implies a huge mass in a very small volume.

Some spirals have Active Galactic Nuclei in their centres, like those in elliptical galaxies that we mentioned in the previous chapter, so although our Galaxy is pretty much inactive (the luminosity of Sgr A* is only $10^3 L_\odot$), it could reasonably be assumed that this mass must be largely concentrated in a central black hole.

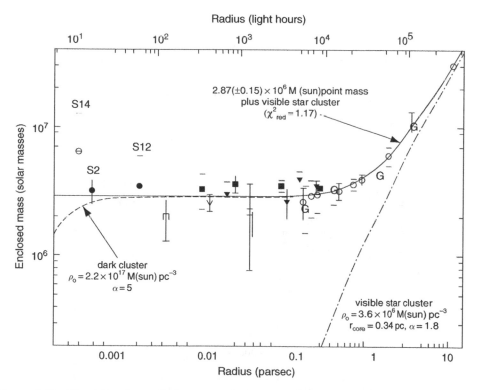

Figure 4.43 The derived cumulative mass as a function of distance from the Galactic Centre (in light hours). The dynamics of the central region are best fitted by the dense visible star cluster plus a black hole of mass $2.9 \times 10^6 M_\odot$ (Reproduced with permission from Schödel *et al.*, 2003).

Remarkably, we can go a step further towards proving this, by actually watching individual stars orbit around Sgr A*. Using observations taken over a span of about 10 years, it has been shown that a star which goes by the prosaic name of S2 is orbiting with a period of 15.2 years and at pericentre was just 120 AU from Sgr A* and moving at 8000 km s^{-1}. A trivial application of Kepler's laws gives a central mass of $3.3 \times 10^6 M_\odot$. Allowing for the known stellar cusp, all the observational data are well fitted by a model with a $(2.9 \pm 0.1) \times 10^6 M_\odot$ black hole in the centre. Even without the black hole, conditions would be pretty bizarre, with a stellar mass density near the centre of around $10^8 M_\odot$ pc^{-3}, and physical stellar collisions must be commonplace.

The other local spirals have somewhat different central regions. In high resolution imaging, M31 is seen to have a double nucleus. One component appears to contain a $10^6 M_\odot$ black hole, the other, about 2 pc away, to be a compact star cluster. Interestingly, though, the very central regions appear to contain no gas or dust. The dynamics of the nuclear region of M33 shows no evidence for a black hole, though it does have a strong central X-ray source. It also possesses a nuclear star cluster,

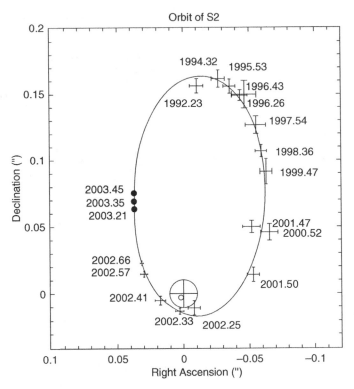

Figure 4.44 The orbit of the star S2 around the Galactic Centre (Reproduced with permission from Eisenhauer *et al.*, 2003).

comparable to a large globular cluster with luminosity $\sim 2 \times 10^6 L_\odot$, but with M/L only around 0.4. These nuclear clusters appear to be common in late type spirals and also appear in some dwarf galaxies. Given the low M/L (at least for the ones in spirals), they may be associated with episodes of star formation. Abundant gas in the central 100 pc is a common feature of spirals of all types.

5 Irregulars, dwarfs and LSBGs

Once we leave the realm of the giants, we encounter a rather more confusing and often relatively ignored component of the overall galaxy population. Commenting in his 1943 book on his recent discovery of dwarf spheroidal galaxies, Shapley aptly summarised the impact they had had on galaxy studies as 'Two hazy patches have put us in a fog.' In some ways, despite the hope that small systems should be somehow 'simpler', this fog has yet to lift.

Some galaxies are simultaneously members of all three of the title categories, being irregular in morphology, low in luminosity and of low surface brightness, while many have two and some just one of these characteristics. In general terms, irregular galaxies are those with star formation but no spiral features. Some can be quite luminous, but most are dwarfs (of type dI). Dwarfs are, by definition, of low luminosity and usually (but not necessarily) also have small sizes. They can either be irregular or smooth in appearance, the latter being the non-star-forming dEs or the even lower luminosity dSphs. Many dwarfs are also of low surface brightness, but there do exist high surface brightness dwarfs, such as the BCDs first identified by Zwicky in the 1950s. Apart from possessing their eponymous defining characteristic – specifically a low *central* surface brightness (as most galaxies are of low surface brightness at large radii) – LSBGs can have many forms. Both dwarf ellipticals and dwarf irregulars can be (and often are) of low surface brightness, and all dwarf spheroidals qualify. However, some large disc galaxies also have surprisingly low surface brightness and the largest known disc galaxy, Malin 1, which has a scale size of about 80 kpc is also the giant galaxy with the lowest central surface brightness, around 26 Bμ or just $2L_\odot \, \mathrm{pc}^{-2}$.

The Structure and Evolution of Galaxies   Steven Phillipps
© 2005 John Wiley & Sons, Ltd

5.1   Local Group members

For obvious reasons, the lowest luminosity galaxies are visible only at small distances. Fortunately, the Local Group contains numerous examples; indeed, the large majority of its members are dwarfs of one sort or another.[1]

Dwarf galaxies have traditionally been taken to be those with absolute B magnitudes fainter than $-16 + 5\log h$, where h is the normalised Hubble parameter we met in section 2.6. For its currently likely value of 0.7, this would put the boundary at $M_B \simeq -16.8$; for simplicity we can choose -17 instead. This corresponds to a luminosity $1 \times 10^9 L_\odot$. Remember that distances in the Local Group (LG) are not calculated from recession velocities, so the luminosities of LG dwarfs do not themselves depend on our choice of Hubble's constant.

With this division, the LG contains just four giants (the Galaxy, M31, M33 and, marginally, the LMC), all of them spirals (with again the LMC only just making it, with a classification of Sm). The next brightest LG object is not the SMC, the brightest true irregular at about $3 \times 10^8 L_\odot$, but NGC 205. The latter is the brightest

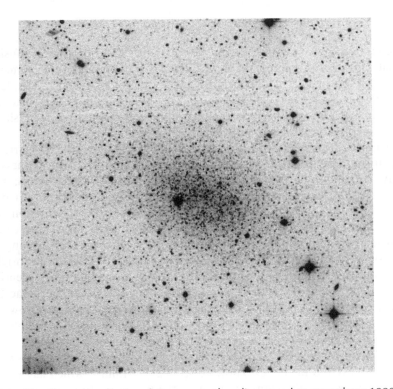

Figure 5.1   The Cetus dwarf, a Local Group member discovered as recently as 1999. (Image courtesy Mike Irwin and the Isaac Newton Group.)

---

[1] This celestial dominance seems appropriate given that in Saxon and Norse mythology the sky was held up by dwarfs!

of the dE satellites of M31 and has a luminosity $\simeq 4 \times 10^8 L_\odot$. The rather strange high surface brightness small elliptical M32, an even closer companion of M31, also has a comparable luminosity. There are then two further largish irregulars, NGC 6822 and IC10, which are not close companions of any of the giants, and two more dEs which are, NGC 147 and NGC 185, both neighbours of M31. Each of these just exceeds $10^8 L_\odot$ ($M_B \sim -14.5$). Below this we find a large number ($\simeq 25$) of dI and dSph systems with magnitudes extending down to $M_B \sim -8$. With one or two possible exceptions, the more numerous dSphs are satellites of the Galaxy or M31 while the dIs are all 'free flying' group members. New members of the Local Group, and indeed satellites of our Galaxy, continue to be found even today.[2] We will return to the case of one of these recent discoveries in section 5.4.

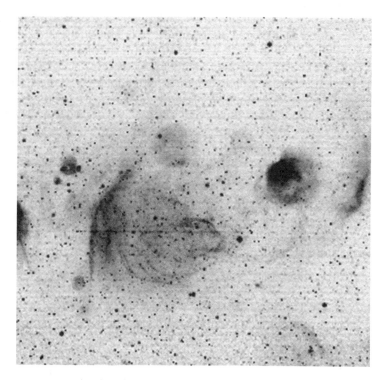

Figure 5.2   An Hα narrow-band image of the SMC, illustrating the complex nature of the (ionised) ISM. (Courtesy: CTIO)

5.2   Irregulars

While the LMC is an example of a galaxy at the border between the late type spirals and irregulars – basically disc-shaped though with a strong bar – the SMC is almost

---

[2] For some time the Galaxy had the correct fairy-tale number of seven dwarfs as companions, but now there are nine or ten.

all bar with no disc to speak of. It points roughly towards us and has a depth of about 15 kpc along the line of sight (so not particularly dwarfish in that sense). Both Clouds are gas rich, with HI masses around $6-7 \times 10^8 M_\odot$ and are typical of irregulars and dwarf irregulars in having $M_H/L_B \simeq 1/2$ to 2. Each has numerous star-forming regions, with the LMC containing the huge 30 Doradus complex, just off the end of its bar. Again this is characteristic of irregulars, the strong current star formation lending them their very blue colours as well as their optical morphologies. The ISM in irregulars, as exemplified by the Magellanic Clouds, is very complex, with many loops and filamentary structures, numerous cavities apparently having been blown in the ISM by stellar winds and supernovae. In some cases, like NGC 6822 and IC10, the distribution of the neutral hydrogen in irregulars can be up to ten times more extended than the optical light.

Photometrically, irregulars usually show underlying exponential profiles, which are more obvious at red or infra-red wavelengths, where the effect of the most recent clumps of star formation is reduced. The LMC and SMC both have similar exponential scale sizes, 1.5 kpc or a little less, compared to the 4–5 kpc for the Galaxy and M31. In addition, the broad, fairly flat distribution of observed axis ratios b/a demonstrates that, notwithstanding the case of the SMC, irregulars are generally disc-like rather than ellipsoidal (though they are not as flattened as spiral discs).

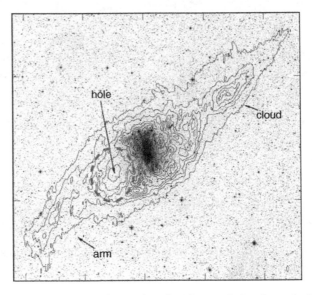

Figure 5.3 HI map of NGC 6822 (Reproduced with permission from de Blok and Walter, 2000) showing the large extent of the gas (contours) relative to the starlight (greyscale).

Although not simple to determine in some cases, the disc central surface brightnesses appear to correlate with luminosity, at least for moderately bright systems. This is used in the ImIII to ImV classification for irregulars which continues the luminosity class idea, surface brightness being used as the luminosity indicator in place of spiral structure. However, at lower luminosities, even this rough correlation

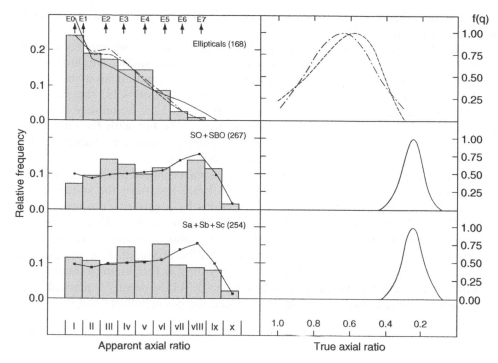

Figure 5.4 The b/a distributions for different galaxy types (Reproduced with permission from Sandage, Freeman and Stokes, 1970). Dwarf irregulars show a broad distribution similar to that of the spiral discs while dwarf ellipticals follow that of giant ellipticals.

appears to break down. Local group dIs, typical of their class, have central surface brightnesses in the range 21 Bμ down to about 25 Bμ.

The star formation in irregulars is clearly not driven by a spiral density wave and, with the exception of the SMC, none of the LG irregulars has a close companion. Indeed, the lack of any real organisation suggests a completely different mechanism, perhaps that called 'stochastic self-propagating star formation'. In this picture, star formation in one area leads to supernovae and mass outflow which then sweeps up material as it propagates into neighbouring parts of the ISM, triggering further cloud collapse and a new burst of star formation. In this way the star formation can propagate around the galaxy, but cannot quickly return to a previous star-forming area as the fuel there has been used up. In the LMC, for example, the luminous star formation region R136 appears to have been triggered by star formation elsewhere in the 30 Dor region.

5.2.1 Metallicities

Irregulars show generally low metallicities, even in recently formed stars and the ISM, typically 1/10 solar or less. If we determine Z from the emission line spectra of their HII regions we find that the current metallicity is correlated with the overall luminosity, the smaller systems having lower Z. At least partly as a consequence of their low metallicities, dIs contain relatively little obscuring dust and are therefore not very bright in the FIR.

Notice that since surface brightness declines with luminosity, we might also argue that the important correlation is between metallicity and surface brightness. For example, it is the surface density which drives the star formation, so it should also influence the increase in metallicity. As we saw in section 4.13, the gas-phase metallicity will depend on the remaining gas fraction. Low luminosity irregulars certainly do have high gas fractions (say of order 1/2), but with the closed box models this would generate $Z \sim p \ln 2 \sim Z_\odot/2$. To obtain much lower values we have to appeal to models with gas flows, in this case outflows.

Strictly, we should modify our equations of chemical evolution to allow both for the time dependence of gas turning into stars and being lost from the system in the outflow. However, we can get an idea of the effects by considering a simpler model where the star formation ceases when the ISM becomes too diffuse. We could imagine this occurring if the ISM is liable to be swept out of a galaxy if its surface density drops below a critical level, Σ_c. An equivalent effect will be produced if, as discussed in section 4.12, there is a minimum threshold density for star formation to occur.

Suppose the stellar surface density when this critical ISM density is reached is Σ_s. We know from the equations in section 4.13 that the gas metallicity will be

$$Z = p \, \ln(1 + \Sigma_s/\Sigma_c) \tag{5.2.1}$$

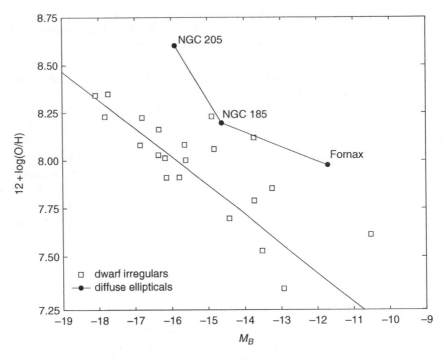

Figure 5.5 The luminosity–metallicity relation for dwarf galaxies (Reproduced with permission from Richer and McCall, 1995, ApJ). Note the offset between the dE/dSph galaxies (circles) and dI galaxies (squares).

and the mean metallicity of the stars at this point must be

$$<Z_*> = p\left[1 - \frac{\Sigma_c}{\Sigma_s}\ln\left(1 + \frac{\Sigma_s}{\Sigma_c}\right)\right] \tag{5.2.2}$$

If there is originally a large mass of material then most of the gas can be used up before its density drops to Σ_c. Thus the mean metallicity tends to p as for the ordinary closed box model. This will then be the case for galaxies with large stellar masses and hence luminosities. However, if only half the gas is used up before the remainder is lost or becomes too diffuse for further star formation activity, then $<Z_*> \simeq 0.3p$. If the initial density were so low that $\Sigma_s \simeq 0.1\Sigma_c$, then $<Z_*>$ drops to only $0.05\,p$ or about $Z_\odot/30$, and the metallicity in any remaining gas will be about twice this. This might apply to very low mass objects which very quickly lose the majority of their gas. This mechanism automatically implies a correlation between surface brightness and metallicity, of course.

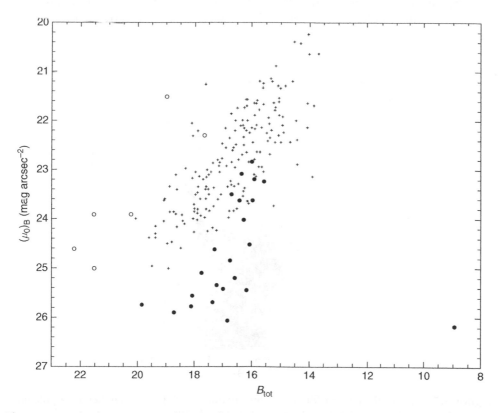

Figure 5.6 The luminosity–surface brightness correlation for dwarf galaxies in the Virgo Cluster (Reproduced with permission from Bothun, Impey and Malin, 1988).

5.3   Early type dwarfs

The most plentiful type of dwarf in the Local Group, and probably the most common variety of galaxy in the universe is the dE, along with its extension to even lower luminosity, the dSph. They are particularly numerous in rich clusters like Virgo and are often said to be the most clustered galaxies of all, seldom if ever appearing except in proximity to giant galaxies. The dSph galaxies are the lowest luminosity galaxies known, with L as low as $2 \times 10^5 L_\odot$, less than that of the largest globular clusters and comparable to the brightness of a single main-sequence O5 star! And IX, a satellite of M31 discovered in 2004, appears to have taken the record for faintness at $M_V \simeq -8.3$, just beating the Draco and Ursa Minor dSph satellites of our Galaxy.

Interestingly dEs and dSphs have intensity profile shapes which are close to exponential as for the dIs, though their axis ratio distribution shows clearly that they really are spheroidal in shape, rather than disc-like. As also noted in section 3.2, there is evidence that the profile shape is dependent on luminosity, with the larger dEs having profiles more like giant ellipticals and the more closely exponential shapes only appearing at lower luminosities. A significant fraction of dEs – perhaps the majority of the brighter ones – contain central nuclear star clusters, somewhat like the one in M33 mentioned in the previous chapter. These nuclei have luminosities comparable to globular clusters and still appear, though less frequently, in quite faint hosts.

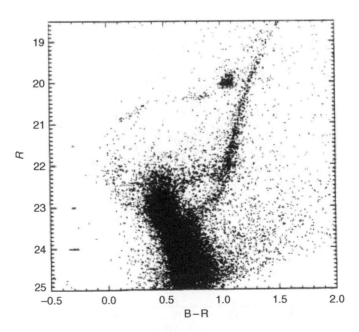

Figure 5.7   The H–R diagram for the Carina dSph, showing multiple stellar populations (Reproduced with permission from Smecker–Hane *et al.*, 1996).

The dSphs have the lowest surface brightnesses of all dwarf galaxies, explaining why the first examples of the class, the Fornax and Sculptor dwarfs, were discovered by Shapley only in 1938. In fact, the nearby examples in the LG are detectable as an excess of individual faint stars, not via their averaged out surface brightness, which was below detectable levels until much recently. Though the smallest dSphs are not much more luminous than the Galaxy's brighter globular clusters (indeed the famous globular ω Cen is *more* luminous than about 10 of the faintest spheroidals) and they have similar stellar populations and old ages, they are quite distinct in other ways.

For a start, they are much larger. They have half light radii in excess of 200 pc compared to only $\sim$5 pc for globulars, and so are correspondingly much more diffuse. Their central surface brightnesses (determined by counting the number of visible stars per unit area) can be as low as about 26–27 Bμ, that is, just one or two solar luminosity stars per square parsec, with correspondingly low volume densities, below $0.01 L_\odot$ pc^{-3}. Globular clusters, by contrast, have stellar volume densities of 10^2 to 10^4 stars per cubic parsec and follow nearly isothermal radial density laws called King profiles. In addition, dSphs formed their stars over a much more extended period (of order Gyr) than did (nearly all) globulars. For instance, even the tiny Carina dSph shows very clearly in its H–R diagram the evidence for separate main-sequence turn-offs at ages around 13, 6 and 3 Gyr. During this time the metallicity gradually increased, the resultant metallicity spread again distinguishing dSph galaxies from (nearly all) globular clusters. (The *caveat* is that a few large globulars, like the aforementioned ω Cen, *do* have more complex stellar populations than their smaller brethren.)

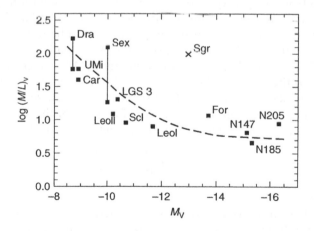

Figure 5.8 Compilation of M/L ratios of early type dwarfs (Reproduced with permission from Mateo, 1997).

Also, while globulars have $M/L \simeq 1$–2 (indicating that they are formed entirely of stars), dE and dSph galaxies can have huge values of M/L, judging by the surprisingly high velocity dispersions of their individual stars. The deduced M/L can be anything from around 10 to approaching 100, with the smallest galaxies, such as Draco (which has a scale size of only 140 pc), apparently the most dark matter dominated. Indeed, the increasing M/L for the lower L systems means that all the faint dSph in the LG have rather similar masses, from 1×10^7 to $3 \times 10^7 M_\odot$. The rotation velocities of dEs are often (but not always) very small, their shapes being produced by anisotropy of the velocity dispersion as for giant ellipticals.

The early type dwarfs generally show no sign of recent star formation. They are also pretty much free of interstellar material (even the very hot gas seen in giant ellipticals). However, there are some exceptions to this, for instance the central regions of NGC 205 and NGC 185 contain up to $10^6 M_\odot$ of gas and young stars less than about 500-Myr-old.

5.3.1   Galactic winds

Like dIs, both dEs and dSphs are metal poor, with Z down to about $1/15 Z_\odot$. (The lowest metallicity galaxy known, the blue compact dwarf IZw18, has $[O/H] = -1.7$, corresponding to $1/50 Z_\odot$.) The dEs also demonstrate the correlation we noted for dIs between luminosity and metallicity, though this is now the mean metallicity of their stars, of course, not a gas metallicity. The L–Z relationship for the early type dwarfs has nearly the same slope but is offset from that of the irregulars (Figure 5.5) and also extends the relationship observed for early type giants. Recall from Chapter 3 that the overall trend is believed to be due to the shallower potential wells of the lower luminosity, lower mass objects. Once a generation of stars has formed, supernovae will quickly result and can drive galaxy scale winds. If the mass of the galaxy is small, these winds can easily attain escape velocity, so that the newly enriched gas is swiftly removed from the galaxy before much further star formation can take place.

This same process may be responsible for the decrease in surface brightness with decreasing luminosity amongst dwarf ellipticals and spheroidals. Consider, for instance, the following simplified version of a model originally proposed by Avishai Dekel and Joe Silk in 1986. The energy input into the ISM by supernovae E_{SN} should be proportional to the number of supernovae which have occurred and therefore to the mass of stars which have formed, M_s. If the stars constitute some fraction s of the baryonic mass (and the baryons are some fixed fraction b of the total halo mass M), then the binding energy of the baryons in the potential well of the halo will be

$$E_{\text{bind}} \propto bM\frac{M}{R} \propto \frac{M_s}{s}\frac{M}{R} \tag{5.3.1}$$

where R is the size of the galaxy.

Now assume that all galaxies (or, more accurately, their dark halos) are formed with similar overall mass densities (a not-so-unreasonable approximation), so $M \propto R^3$. Thus $M/R \propto M^{2/3} \propto M_s^{2/3} s^{-2/3}$ and the binding energy becomes

$$E_{\text{bind}} \propto M_s^{5/3} s^{-5/3} \tag{5.3.2}$$

Thus E_{SN} will overcome the gravitational binding of the ISM, and hence drive the gas out of the galaxy, when s reaches some critical value which satisfies

$$M_s \propto M_s^{5/3} s^{-5/3}$$

i.e.

$$s \propto M_s^{2/5} \tag{5.3.3}$$

In other words, the star formation will be cut off at a lower stellar fraction, and therefore higher remaining gas fraction, in galaxies with low stellar masses. Evidently, the surface brightness then becomes $I \propto M_s/R^2$ (we can assume that the stellar system has a radius proportional to the halo it inhabits), or

$$I \propto M_s M^{-2/3} \propto s^{2/3} M_s^{1/3} \propto M_s^{3/5} \tag{5.3.4}$$

giving us the desired decrease in surface brightness with decreasing stellar mass and hence luminosity.

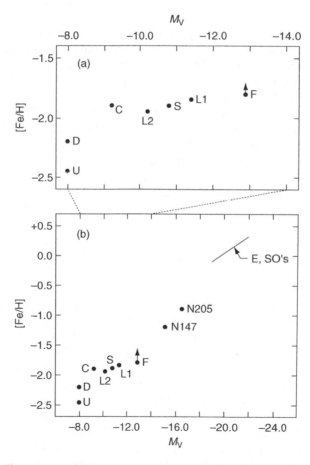

Figure 5.9   Metallicity versus luminosity relation for LG dSph (Reproduced with permission from Aaronson and Mould, 1985). The lower panel shows a compilation for various galaxy types.

In addition, since we know that the metallicity is lower for galaxies with high gas fraction, this obviously also results in low metallicity for low luminosity galaxies. When a (low) fraction s of the baryons has been turned into stars we can make the approximation $Z = p \ln (1/(1 - s)) \simeq ps$. Thus from above

$$Z \propto s \propto M_s^{2/5} \propto L^{2/5} \qquad (5.3.5)$$

in rather good agreement with observation.

5.4   Star formation histories

The Magellanic Clouds contain large numbers of star clusters. Besides a few old ($\simeq 11$ Gyr) globular clusters, the LMC, in particular, hosts many much larger open clusters than does the Galaxy. They present an appearance somewhat like the young globular clusters discussed in section 3.9 in the context of interacting galaxies. Almost all have ages less than 3.5 Gyr, suggesting a long hiatus in star formation. However, the ages then range all the way down to a few Myr; the most luminous cluster, R136 in 30 Dor, is only 3.5-Myr-old and has $L \simeq 10^7 L_\odot$. Observations of field stars show a similar dramatic rise in the SFR over the last 3 Gyr to the clusters, possibly with a peak 1 Gyr ago. There is no corresponding gap in SMC cluster ages, though, with the bulk of its stars in the age range 12 down to 3 Gyr. It has much less prominent HII regions, consistent with a quite low current SFR. Most of the stars less than 1-Gyr-old appear to be in the 'Wing' to one side of the main body of the galaxy and in the central bar-like feature.

Star formation histories (SFHs) of other local dwarf galaxies, that is, the variation of their SFR with time, can nowadays be investigated in some detail by using observations with the HST to place large numbers of individual stars in the H–R diagram. The density of stars in different parts of the diagram can be interpreted in terms of the number of stars of different ages, a more sophisticated version of looking for multiple main-sequence turn-offs.

The most striking thing about these recent HST results is that there seems to be no such thing as a typical dwarf. Each LG dwarf has its own individual history. For instance, Draco is very old, even when compared to globular clusters such as M92, while as we saw above, Carina formed its stars in three well-separated bursts. Fornax also clearly shows distinct old, intermediate age and young stars. NGC 147 contains nearly all old stars and clearly started its main star formation more than 10 Gyr ago, while the otherwise very similar NGC 205 was actively forming stars only 800 Myr ago, and continues to do so in its nucleus. Phoenix, classified as a borderline dSph/dI has a dominant old population but with some stars less than 10-Myr-old. Among the irregulars, while they are all forming stars now, some like NGC 6822 clearly formed stars around 12 Gyr ago (it even has at least one globular cluster) and are currently forming them at only a very low rate (around $0.02 M_\odot$ per yr). IC10, conversely, is in a burst of star formation at present and at the current rate will consume all its gas in

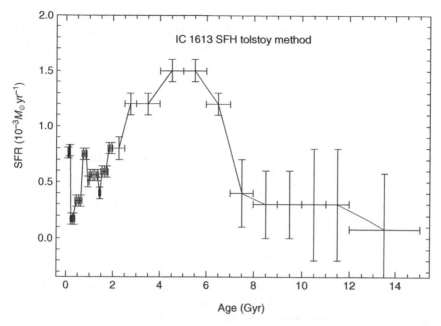

Figure 5.10   The star formation history of IC1613, showing a star formation episode about 3 to 6 Gyr ago (Reproduced with permission from Skillman *et al.*, 2003).

a couple of Gyr. Pegasus and Leo A formed most of their stars 2–4 and 1–2 Gyr ago respectively.

A plausible reason for the variation is close interaction with the Galaxy or M31 at various times in the galaxies' histories. Gravitational effects during a close encounter may lead to compression of the gas in the dwarf and hence initiate collapse of gas clouds and star formation. It is worth noting that dwarf galaxies, generally, do not seem to possess the dense molecular gas out of which spiral galaxies manufacture their stars, so dwarfs appear to be able to form stars via alternative routes to giants.

5.5   Interactions

There are many indications of past interactions in the Galaxy–LMC–SMC system. First, there is the 'bridge' of gas – also containing some quite young star clusters – between the two Clouds. Secondly, there is a more tenuous trail of gas clouds ahead of and behind the SMC in its orbit, called the Magellanic Stream. It extends nearly half way around the sky and contains around $2 \times 10^8 M_{\odot}$ of HI (comparable to the amount in the Clouds themselves). The orbits of the Clouds are highly eccentric, bringing them very close to the Galaxy every 2 Gyr or so. The last close approaches were ∼200–400 Myr ago. Currently the two Clouds are about 20 kpc apart and ∼50 kpc from the centre of the Galaxy. However, at their closest approach to each other (only some 50 Myr ago), their separation was probably only 10 kpc, implying a very close or even interpenetrating collision. The SMC has clearly been severely affected by tidal forces and may now be hardly gravitationally bound at all. In addition, as we saw in Chapter 3, tidal friction from the Galaxy will bring about the decay of the Clouds' orbits and eventually lead to their merger with the Galaxy itself.

This brings us neatly to the case of the most recently discovered of the Galaxy's close satellites. In 1994, while examining redshift data for stars in the direction of the Galactic bulge, Rodrigo Ibata, Gerry Gilmore and Mike Irwin discovered a large group of stars all evidently moving with closely similar velocities, but very different ones to those expected for bulge stars. Further work showed that these stars were part of a satellite galaxy – the Sagittarius dwarf – currently plunging through the opposite side of the Galactic Plane to where the Sun lies and only about 16 kpc from the centre of the Galaxy. Stars pulled out of the dwarf are found in a stream more than 20° long, corresponding to at least 10 kpc. Oddly enough, part of the Sagittarius dwarf has been known since the 18th century. The supposed Galactic globular cluster M54 is certainly associated with the Sagittarius dwarf and may even be its nucleus. At least three, and possibly five other, globular clusters also appear to belong to the galaxy, suggesting that it was once much more massive than its current self. Only Fornax of the other LG dSph galaxies has any globular clusters. The Galaxy may have accumulated other globulars in the same way, from previously disrupted satellites. Indeed, over the last few years, a number of 'streams' of stars have been found around the Galaxy (and around M31), suggestive of the fossil remains of completely accreted dwarf galaxies supplying a substantial fraction of halo stars. One stream around the Galactic Plane seems to be associated with an over-density of infra-red detected stars in the direction of Canis Major which is probably the remnant of one such, now more or less completely destroyed, dwarf companion.

Even when not actually interacting in a collisional sense, satellite galaxies are still subject to gravitational tidal effects from their giant neighbours. If we can approximate the satellite as a point mass m orbiting a distance d from the giant (which has mass M), then a star will remain bound to the satellite only if it is inside the Roche limit radius

$$r \simeq \left(\frac{m}{3M}\right)^{1/3} d \qquad (5.5.1)$$

This limit arises because there is a lower gravitational acceleration due to the primary, for a star on the far side of the satellite relative to the primary, than there is at the satellite's centre, by an amount

$$\frac{GM}{d^2} - \frac{GM}{(d+r)^2} \simeq \frac{2GMr}{d^3} \qquad (5.5.2)$$

However, the required acceleration to keep both parts of the satellite orbiting with the same angular velocity about the primary, $\omega = (GM/d^3)^{1/2}$, is actually greater for the point on the far side by

$$\omega^2(d+r) - \omega^2 d = \omega^2 r = \frac{GMr}{d^3} \qquad (5.5.3)$$

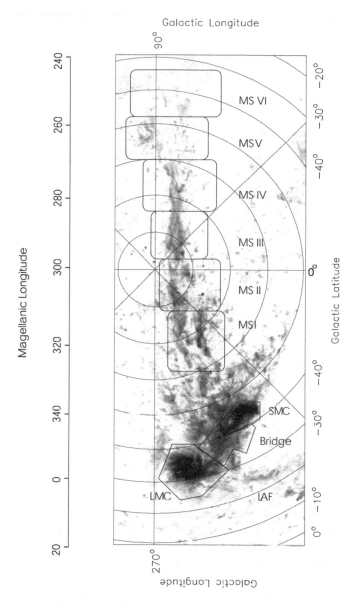

Figure 5.11 HI map of the Magellanic Stream (Reproduced with permission from Putman *et al.*, 2003).

Thus the star will be lost unless the combined shortfall $3GMr/d^3$ is less than the satellite's self-gravity Gm/r^2, leading to the criterion stated above. This is also essentially the criterion that the star orbits its own galaxy in less than the orbital time of the satellite around the giant galaxy.

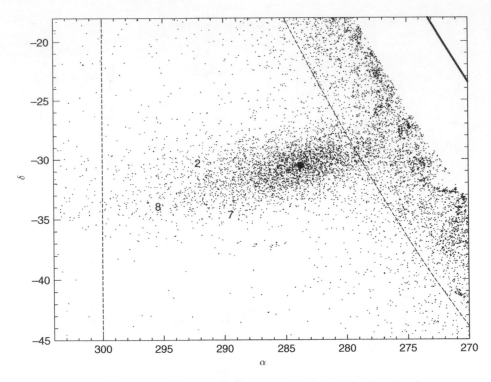

Figure 5.12 A plot of the M giant star density towards Sagittarius, showing the stars associated with the Sagittarius dwarf as a plume perpendicular to the Galactic Plane (Reproduced with permission from Majewski *et al.*, 2003).

Effects such as these may have been important in M32. Though it is sometimes put forward as a direct low mass extension of the giant elliptical sequence, continuing the trend for higher surface brightness with decreasing luminosity, it seems more likely that it is a rather distorted version of a higher mass galaxy. Sometimes referred to as a compact dwarf elliptical, it has a central surface brightness 10 magnitudes higher than typical dwarf ellipticals and has the metallicity expected of a much more massive galaxy. The dynamics of its central regions show evidence for a supermassive black hole of $\simeq 3 \times 10^6 M_\odot$ and the galaxy as a whole shows significant rotation. Hence it may reasonably be interpreted as the remnant of a galaxy, perhaps of an early type spiral, which has lost its outer regions due to the tidal effects of M31. M32 also possesses no globular clusters. It has been estimated that its orbital period about M31 may be around 800 Myr and its orbital radius only of order 12 kpc. If so, like the Sagittarius dwarf in our Galaxy, M32 will periodically plunge through the outer disc of M31.

A possibly related group of objects are the UCDs, the most recently added members of the galaxian zoo, first identified in 2000. These have luminosities comparable to other dwarf galaxies but tiny effective radii of just 15–20 pc. This

FCC303 dE, N **UCD 0339–3504**

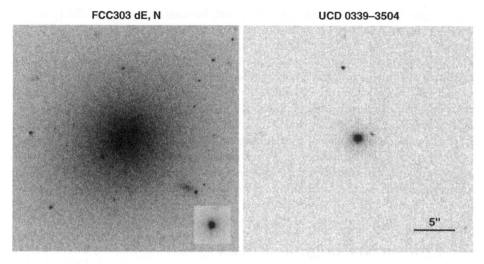

Figure 5.13 HST images of a dE,N galaxy (left) and a UCD (right), emphasising the similarity in size of the UCD to the dE nucleus (inset). (Courtesy: M. Drinkwater *et al.* and NASA).

means that even though the known examples are quite nearby they can be studied in detail only by use of the HST; in ordinary ground-based telescope images they look just like stars and they were found only because they had the (redshifted) spectra of galaxies. In many ways they appear to be transition objects between galaxies and globular clusters, but their origin may lie in larger galaxies. Only detected to date in the Virgo and Fornax clusters, they may be the remnant nuclei of former nucleated dwarf elliptical (dE, N) or late type spiral galaxies which have been tidally stripped of their outer parts as they fly through the central regions of their clusters.

Finally, while considering the mix of dwarf types, we might note that there is also evidence for the *formation* as well as destruction of dwarfs via interactions. The so-called 'tidal dwarfs' appear as dense knots of gas in tidal tails, and also contain young stars, indicating that they are condensing out of the tail gas into galaxies in their own right.

5.6   Interconnections

In an attempt to make sense of the numerous observed types of dwarfs, it has been natural to look for evolutionary connections. Besides the destructive processes invoked above, one obvious possibility is that once a BCD or dI runs out of gas the star formation is cut-off and the galaxy gradually fades, becoming redder in the process. This might naturally lead to the production of a dE or dSph galaxy. However, a difficulty with this simple picture is that some dE galaxies appear to have *higher* surface brightnesses than similar-sized dIs, clearly contrary to the fading picture. Nor is there any convincing evidence for lower surface brightness objects being progressively redder, though this could be confused by metallicity effects. A variant which might overcome some of the difficulties sees the dI evolve via a final burst of star formation which temporarily elevates it to higher surface brightness – perhaps even as a BCD – before the final fading to an early type dwarf. On the other hand, there may be detailed problems with this scheme in terms of the different element abundances seen in dE galaxies and, at least, the strongly star-forming objects like BCDs, which suggest that dEs or dSphs had much more continuous low level star formation. Nevertheless, it is hard to see what else a dI could look like once its star formation ceases and there are examples of galaxies which appear to be transition objects, for instance the Pisces dwarf (also known by its original name LGS3, for 'Local Group Suspect'). However, one would have to admit that there is currently no clear picture of the interconnection between the various dwarf types.

5.7   Low surface brightness discs

Besides the dwarf galaxies with low surface brightness, there is a more recently appreciated class of larger disc galaxies which also have low surface brightness. Some authors specifically imply this type of object when they say 'low surface brightness galaxy'. Others, and we will, continue to follow this usage, consider any galaxy, of any size, which has a low central surface brightness to be an LSBG. The large low surface brightness (LSB) discs have come to light mainly through technological improvements. In section 4.1 we noted that the use of photographic material with a relatively bright isophotal limit could bias us towards or against certain types of galaxy, and this prediction was borne out when, for instance, POSS-II, produced in the late 1980s, superseded the original Palomar survey carried out in the 1950s. Many more LSB discs were discovered and some authors have gone so far as to propose that they form a third parallel sequence alongside normal and barred spirals – making a trident instead of a tuning fork – with similar frequencies to the other forms.

The importance of this population remains controversial, however, other workers arguing that luminous LSB disc galaxies are quite rare. To some extent, both views may be correct, since we can have equal numbers of normal and low surface brightness discs at any given *scale size*, yet few LSB discs with high luminosity, since the latter would require exceptionally large sizes to make up for their low surface intensities.

The most extreme LSB disc is still that of the galaxy Malin 1, discovered in 1987 with the aid of photographic image enhancement. It has a scale size of around 80 kpc, giving it a total luminosity above L_*, and a huge HI mass, but on unprocessed sky survey images only its small bulge was visible.[3] The central surface brightness of its disc is only about 26 Bμ, and it and its 'cousins' appear to be turning

---

[3] It was originally thought to be a small galaxy in the Virgo Cluster, but its HI spectrum showed it to be much more distant.

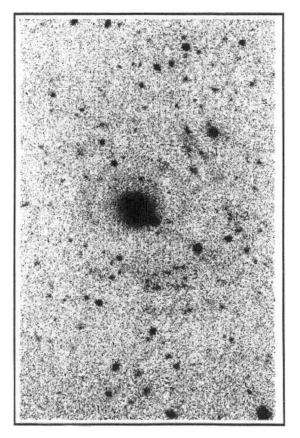

Figure 5.14 The extreme LSB disc galaxy Malin 1 (Reproduced with permission from Bothun, Impey, Malin and Mould, 1987).

their gas into stars at a particularly low rate (per unit area). They also lack molecular clouds.

Large LSB discs rotate like their more prominent counterparts. Indeed they fit on essentially the same Tully-Fisher relation. Their rotation curves indicate that they have large M/L and are dominated by dark matter even close to the centre, *unlike* the case for bright spirals. However, this very dominance has led to a possible problem for the dark matter models, as the deduced mass density profile does not appear to fit that expected theoretically for the dark matter halos.

5.8 Numbers and selection effects

The whole question of the number of dwarf and LSB galaxies and their contribution to the LF is fraught with difficulty, because of the strong selection effects against

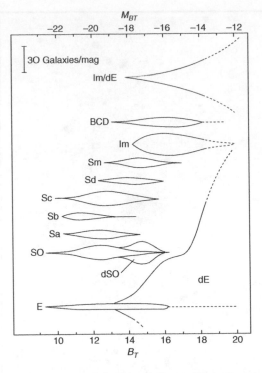

Figure 5.15 A schematic of the LF of the Virgo Cluster (Reproduced with permission from Sandage, Binggeli and Tammann, 1985), showing the contribution from each galaxy type. The width of each figure represents the number of galaxies at that magnitude.

including them in magnitude or angular size limited surveys. Recall from section 4.1 that for fixed L the isophotal size of a galaxy with an exponential profile follows

$$R_{\mathrm{lim}} \propto I_0^{-1/2} \ln(I_0/I_{\mathrm{lim}}) \propto 10^{0.2(\mu_0 - \mu_{\mathrm{lim}})}(\mu_{\mathrm{lim}} - \mu_0) \tag{5.8.1}$$

If we have a survey limited at an isophotal angular size θ_{lim} then evidently such a galaxy will be included, provided that it is nearer than some distance $D = R_{\mathrm{lim}}/\theta_{\mathrm{lim}}$. Assuming a homogeneous distribution of galaxies in space, we will therefore observe a number $N \propto D^3$ of these galaxies. But clearly the distance also depends on the particular value of L as $L^{1/2}$, by the inverse square law. Thus in total we can expect to count a number

$$N \propto L^{3/2} \left[(\mu_{\mathrm{lim}} - \mu_0)\, 10^{0.2(\mu_0 - \mu_{\mathrm{lim}})} \right]^3 \tag{5.8.2}$$

or

$$N \propto (\mu_{\mathrm{lim}} - \mu_0)^3\, 10^{0.6(\mu_0 - \mu_{\mathrm{lim}} - M)} \tag{5.8.3}$$

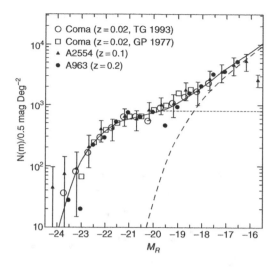

Figure 5.16 The luminosity function of three rich clusters, showing the steep faint end slope (Reproduced with permission from Smith, Driver and Phillipps, 1997).

the actual number we see, depending also on the actual volume density $\phi(L,\mu_0)$, of course. Low luminosity galaxies with low surface brightness are thereby selected against twice over, once through the $L^{3/2}$ term and once through the $(\mu_{lim} - \mu_0)$.

As mentioned earlier, it was originally thought that galaxies had a more or less gaussian-shaped LF. This was primarily because of the neglect of the lower SB dwarfs; the giant ellipticals and normal spirals may indeed have an LF that is quite well approximated by a gaussian. At the faint end, the dwarf galaxy LF shows the power law tail of a Schechter function, but even now the slope of this tail is a matter of considerable debate. One problem is that the overall LF shape is made up of these separate functions for each morphological type, so that what is a good fit at bright and intermediate luminosities may give a misleading impression of the faint end slope. Second, the selection effects noted above mean that there will be relatively few dwarfs in a magnitude or size limited sample, so the statistical uncertainties can be very large. Even worse, depending on the limits of the survey in question, whole ranges of surface brightness may be totally missing from the sample. And last, but by no means least, the LF may be of different shapes in different environments.

Specifically, clusters of galaxies (which should more equally sample the faint galaxies) typically show steeper faint end slopes (often around $\alpha \simeq -1.5$) than do general field samples ($\alpha \simeq -1.1$ or -1.2). There may also be more detailed differences, with the ratio of dwarfs to giants (and hence the LF slope) depending on the local galaxy density. Dwarfs seem to be more widely dispersed in clusters than are the giants, which congregate in the centre, so that the LF becomes steeper in the outer parts of clusters. This might be due to either the destruction of dwarfs in the centres of clusters or the infall of small galaxies onto pre-existing clusters. The mix of different dwarf types is also environment dependent, with dEs, like their giant counterparts, tending to prefer the denser regions. Possible reasons for this are that

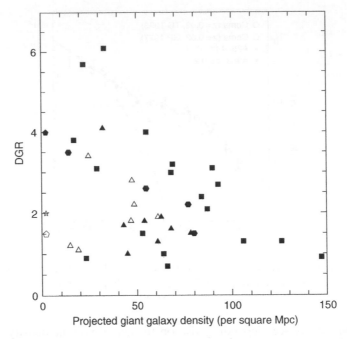

Figure 5.17 The variation of the number of dwarf galaxies per giant (the dwarf-to-giant ratio, DGR) as a function of the local density (Reproduced with permission from Phillipps *et al.*, 1998).

they, too, are the descendants of formerly larger galaxies, which were disrupted as they fell into the cluster centres, or have just spent longer being 'harassed' in the inner regions than the still infalling dIs or spirals.

5.9 Dwarf galaxies in the past

Though we will reserve our main discussion of galaxy evolution until the final chapter, it is also appropriate to mention here the likely variation in the importance of dwarf and irregular galaxies with cosmic epoch. Theoretically, in the popular cold dark matter picture of structure formation, small galaxies should form first, that is earliest, in the development of the universe. Larger systems should then form later by the hierarchical merging of small sub-units. In fact, it was already known before the development of these models, that as we look to fainter and fainter apparent magnitudes, we find more galaxies than can be accounted for by the local galaxy population – the so-called 'faint blue galaxy problem' discussed further in section 8.6. Assuming that these galaxies are at large distances from us, and hence are seen at large look-back times, the 'problem' is the fate of these objects; perhaps they merged with larger systems or faded to near invisibility as their star formation ended and/or their stellar populations aged.

More recently, direct study of the morphology of faint galaxies with the HST, particularly of those in the Hubble Deep Fields, has demonstrated that at high redshifts – corresponding to when the universe was less than half its current age – there were many more irregular galaxies than are seen in the present universe, and that this increase is at the expense of regular well-developed spirals. This has to be tempered by the fact that for galaxies at high redshift, the light we see in the optical was actually emitted in the ultra-violet, where spirals are known to appear less regular due to the greater prominence of individual star forming regions. Nonetheless, there appears to be a distinct shift from the mix of galaxy types we see locally to one where irregulars were much more important. We shall return to this in Chapter 8.

6 Active galaxies

6.1 The discovery of AGN

We have already seen that some galaxies show strong emission lines in their spectra as a result of star formation and emission from hot gas. In 1943 Carl Seyfert produced a small list of galaxies with bright, blue, more or less point-like nuclei in which the emission lines appeared to be far too broad to be due merely to the normal motions of stars or gas clouds within the galaxy: the deduced Doppler widths corresponded to thousands of $km\,s^{-1}$. Most of the galaxies, including the well-known examples NGC 1068[1] and NGC 4151, were spirals, though one was NGC 1275, the central elliptical galaxy in the Perseus Cluster. Further examination showed that the spectra contained lines of a wide range of ionisation states, including, for instance, OVI lines as well as the more usual OII and OIII lines seen in star-forming galaxies.

Although largely ignored for the next 20 years, galaxies showing these characteristics subsequently became known as Seyfert galaxies, the first known 'active galaxies'. Since the broad lines evidently originate in the bright nuclear regions, these are known as 'active galactic nuclei', usually abbreviated to AGN. As we saw when we looked at the Galactic Centre in section 4.17, velocities as large as

---

[1] The unusual spectral lines in NGC 1068 were in fact remarked on by Edward Fath in 1908. Slipher and Hubble also made early observations of galaxies with bright nuclei showing emission lines.

The Structure and Evolution of Galaxies   Steven Phillipps
© 2005 John Wiley & Sons, Ltd

$1000\,\mathrm{km\,s^{-1}}$ occurring close to the centre imply a very large compact mass, most likely a black hole.

The next important surveys in this area were by B.E. Markarian and colleagues at the Byurakan Astrophysical Observatory in Armenia during the 1960s. Utilising 'objective prism spectroscopy' (i.e. by introducing a dispersing prism into the beam before the image was recorded on the photographic plate), they produced lists of galaxies with very blue spectra. Many of these turned out to be additional Seyferts with the characteristic broad-lined spectra. However, as observations became more refined, it became clear that Seyferts needed to be divided into sub-types. Seyfert 1 galaxies were defined to be those in which the hydrogen lines (and other permitted lines like HeII) were very broad, with full widths at half maximum above $1000\,\mathrm{km\,s^{-1}}$, but the forbidden lines[2] like [OIII] were much narrower. The objects in which all the lines had similar Doppler widths of around $1000\,\mathrm{km\,s^{-1}}$ became Seyfert 2s.

Assuming that the line widths are due to the motion of the material near the centre, this implies that in Seyfert 1s (and 1.5 s, 1.8 s and 1.9 s when these further types were interpolated), the permitted and forbidden lines arise in different parts of the overall nuclear region. The lines with the very large widths are therefore ascribed to what is (very reasonably!) termed the Broad Line Region (BLR) and the lines of more modest width to the Narrow Line Region (NLR).

In fact, Seyfert 2s also have a BLR. For example, NGC 1068 is found to show broad lines, but only if viewed in polarised light. This makes sense if its BLR is enshrouded by dust which blocks out the direct light from the BLR (and hence removes the normal broad lines), but scatters some of the light, resulting in polarised broad lines.[3] This discovery was the key to current models of AGN wherein the 'central engine' which generates the energy is surrounded by a very small (parsec-sized) BLR which is itself surrounded by a dust ring or torus. From some angles we can see the BLR directly (Type 1s), but from others the BLR is hidden and its presence is only revealed by its scattered light (Type 2s). Since Seyfert 2s turn out to be more common than Seyfert 1s, this suggests that the torus is quite thick and we see down into the BLR only when viewing from quite close to the axis of the torus. In all, a few per cent of galaxies have Seyfert characteristics, but the distinction becomes quite blurred as lower levels of nuclear activity become detectable.

Going back to the early 1960s, it had also been found that some radio sources, for example from the 3rd Cambridge Catalogue, appeared to be identified with star-like objects. Interferometric observations combine data from multiple radio dishes separated by large distances and can be used to overcome the resolution limit λ/D of a single telescope of diameter D working at long wavelengths λ. Such observations of the source 3C48 showed both that it was very compact (size $<1''$) and pinned down

---

[2] Forbidden lines are those which arise in particularly low density gas but are not seen in normal laboratory conditions. These occur from long-lived or 'metastable' states (lifetimes $\sim 10^4\,$s), so in a dense gas, collisional excitation or de-excitation usually takes place before the atom has time for the relatively slow 'forbidden' transition to occur. Permitted lines are those which do occur rapidly enough (e.g. in $\sim 10^{-8}\,$s for Balmer lines) to be seen under normal conditions. Forbidden lines are denoted by square brackets around the symbol for the relevant ion, for instance [OII].

[3] Emission reflected off dust grains leads to polarisation in the same way that reflection of sunlight off the surface of the sea results in us seeing polarised light.

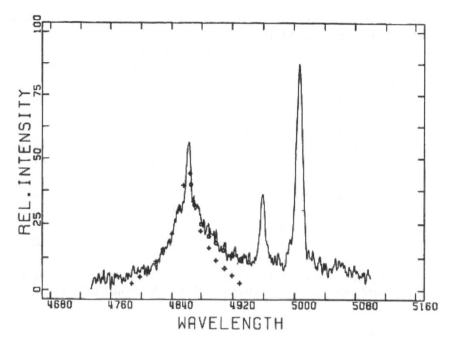

Figure 6.1   The characteristic broad Hβ line in the spectrum of the well-known Seyfert 1 galaxy NGC 4151 (from Capriotti, Foltz and Byard, 1980).

its position to an accuracy of about 5″. This led to its identification with a very blue 16th magnitude stellar object which turned out to have a spectrum containing broad, but unidentified, emission lines.

The breakthrough as to the real identity of these 'radio stars' came following observations of another 3C source, 3C273. This happens to lie in such a position as to be periodically occulted by the Moon. By timing such an occultation in 1962, Cyril Hazard and colleagues at the Parkes radio telescope were able to determine a position accurate to 1″ and show that the source was in fact double. The two components were identified with a 13th magnitude stellar object and a neighbouring elongated nebulosity or jet. Maarten Schmidt at Mount Palomar obtained a spectrum for the stellar object and found that it showed the same puzzling characteristics as 3C48. However, Schmidt then realised that the lines were the standard hydrogen lines but redshifted by a large amount. Indeed the shift would correspond to about $40\,000\,\text{km}\,\text{s}^{-1}$ if it was a simple Doppler shift due to the velocity of the source. This is far larger than could be accommodated within the Galaxy, which has an escape velocity of around $300\,\text{km}\,\text{s}^{-1}$, forcing the conclusion that the 'radio stars' were in fact extragalactic.

Although 3C273 is in fact at relatively modest redshift, $z = 0.16$, other 3C sources were soon found at previously unheard of redshifts, around 2 (e.g. 3C9). At such high z, all the commonly observed optical emission lines are redshifted out into the infra-red and instead we observe lines initially emitted in the ultra-violet, such as the

Lyman lines of hydrogen (particularly Lyα at 121.6 nm), the CIV line at 154.9 nm and, in less extreme cases, the MgII line at 279.8 nm.

But interpreting the redshifts as cosmological implied that the objects – subsequently renamed 'quasi-stellar radio sources' or quasars – were at distances of order 1000 Mpc and must therefore have extraordinary luminosities, above $10^{12}L_\odot$. If these were extremely powerful AGN, it was apparent that in this case the nuclei outshone their host galaxies by at least a factor of 10. Indeed, an oft-used discriminant is to count any AGN with an absolute magnitude brighter than $M_V = -23$ as a quasar. It has only been in relatively recent times, especially since the advent of the HST, that the galaxies underlying quasars have been unambiguously measured at all.

Nevertheless, the size of the emitting region in quasars was quickly established when it was found that many of them were variable on short timescales of a year or less. Except for some contrived circumstances a coherent change in the emissivity on this timescale requires the emitting region to be at most a light year across and probably considerably less (otherwise the differences in light-travel time across the source would 'wash out' the variations).

Subsequently, non-radio-emitting objects with similar spectra and redshifts were identified, primarily through their extremely blue *U–B* colours ('UV excess') com-

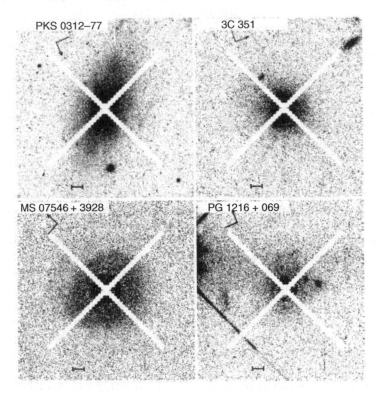

Figure 6.2 Hubble Space Telescope images of quasar host galaxies (Reproduced with permission from Boyce *et al.*, 1998). The central point source (and the corresponding diffraction spikes) have been subtracted off to show the underlying galaxies.

pared to most stars. These became known as just quasi-stellar objects or QSOs. However, with the passage of time, the distinction between the terms 'quasar' and 'QSO' has largely been lost, and both radio-emitting and radio-quiet types tend to be interchangeably known by either name. More recently, other quasars have been discovered via their X-ray emission. This is important because at very high redshifts quasars no longer appear especially blue. Radio-quiet quasars outnumber their radio-emitting counterparts by a factor between 10 and 30.

Originally, perhaps even more enigmatic than the quasars were the BL Lac objects. BL Lacertae, which gives the class its name, had been classified previously as a variable star, varying from 16th to 13th magnitude. Identified with a compact radio source, it showed the same UV excess as quasars but spectroscopy revealed no obvious emission lines from which a redshift could be determined. Later work discerned weak absorption lines from an underlying galaxy component, showing that BL Lac and similar objects were indeed at redshifts similar to quasars. BL Lac objects vary dramatically in brightness, some changing by a factor 2 in luminosity in a matter of hours. Some quasars also demonstrate this trait and are known as 'optically violently variable' or OVV quasars. The OVV quasars and BL Lacs are jointly referred to as 'blazars'.[4] BL Lacs and some radio quasars also show strongly polarised optical emission.

Finally, in this historical section it would be remiss not to acknowledge that some astronomers – led by Halton Arp, Geoff Burbidge and Fred Hoyle – were (and in a few cases still are) unhappy with the cosmological redshift interpretation and

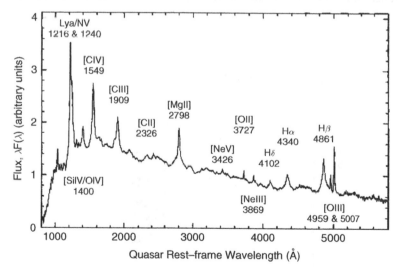

Figure 6.3 The typical QSO spectrum, obtained by averaging over QSOs at different redshifts (Reproduced with permission from Francis *et al.*, 1991), showing the important rest-frame UV and optical lines.

---

[4] The name, appparently, was invented over a conference dinner!

preferred to postulate some sort of intrinsic effect, thus generating many years of lively debate!

6.2   AGN structure

Active nuclei typically emit over a wide range of the electromagnetic spectrum (though as we saw above, many QSOs are radio quiet). Their power output is roughly equal in each decade of frequency, that is $\nu F_\nu \simeq$ constant, right out to the X-ray regime. Indeed in many cases, X-rays in the 0.1 to 20 keV range can make a significant or even dominant contribution to the bolometric luminosity. Since we know from their variability that the huge power output comes from a very small volume, and from the line widths that the mass within this volume must be very large, it is natural to assume that the very central feature of the central engine is a black hole.

According to General Relativity the effective 'size' of a black hole is its Schwarzschild radius

$$R_S = \frac{2GM_{BH}}{c^2} \simeq 3 \times \frac{M_{BH}}{M_\odot} \text{ km} \qquad (6.2.1)$$

the radius of the 'event horizon' from within which no signal can reach the outside universe. As noted in the 18th century by the clergyman amateur astronomer John Michell and mathematician Pierre Laplace, this is the radius at which, in a

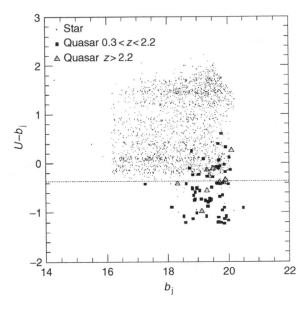

Figure 6.4 A colour–magnitude plot for a sample of point sources, demonstrating the blue U–B colours of quasars (symbols) relative to most ordinary stars (points), the basis of the UVX criterion for QSO selection (Reproduced with permission from Meyer *et al.*, 2001).

straightforward Newtonian interpretation, the escape velocity reaches the speed of light. For a $10^8 M_\odot$ black hole, for instance, it is about 2 AU or 15 light minutes.

6.2.1   Accretion discs

The luminosity we see arises from an 'accretion disc' just outside the Schwarzschild radius (theoretically, the closest stable particle orbit is at $3R_S$, unless the black hole is rotating). Here, infalling material spirals around the black hole (because of its angular momentum), heating up via viscous drag and radiating away some of its potential energy before falling into the hole. In this way, matter can give up around 10% of its rest mass energy (i.e. $\simeq 0.1mc^2$), compared to the $0.007mc^2$ released when a mass m of hydrogen is turned into helium by nuclear processing in stars.

Any magnetic field permeating the gas will be effectively tied to the infalling plasma and is therefore also dragged towards the black hole, increasing its strength. Near the centre it may become strong enough to channel jets of charged particles out along the spin axis of the accretion disc at relativistic velocities. The particles accelerated in this way then produce synchrotron radiation, most usually seen at radio wavelengths. In addition, relativistic electrons can 'up scatter' optical and radio photons right up to γ-ray energies via the 'synchrotron self-Compton' process. X-ray and UV emission (including the so-called 'big blue bump'), which can excite the high ionisation spectral lines, comes from the hottest ($\simeq 30\,000$ K), innermost parts of the accretion disc (or from the jet), with optical radiation arising further out in the disc. Infra-red radiation comes primarily from dust heated by the nuclear emission, the spectrum typically having a peak in the FIR due to dust at distances of order 1 kpc from the centre.

The energy output from the central region can become large enough to be self-limiting, by radiation pressure preventing any further infall. If the nucleus has luminosity L_{nuc}, the flux at radius r is $F = L_{\text{nuc}}/4\pi r^2$. Therefore the rate of momentum transfer across unit area (i.e. the pressure provided) by the photons is F/c (since each photon of energy E has momentum E/c). Electrons in the infalling plasma can scatter any photons within an effective area given by the Thomson cross-section σ_T, so the pressure force on an electron will be

$$F_P = \frac{\sigma_T L_{\text{nuc}}}{4\pi r^2 c} \tag{6.2.2}$$

The maximum L_{nuc} which can be sustained without blowing the plasma out of the system is attained when this just matches the gravitational attraction of the nucleus,

$$F_G = \frac{GM_{\text{BH}} m_{\text{p}}}{r^2} \tag{6.2.3}$$

Notice that this is the force on the *protons*, whereas we have calculated the radiation pressure on the electrons. The corresponding force on the other component is negligible in comparison, but the protons and electrons are strongly coupled

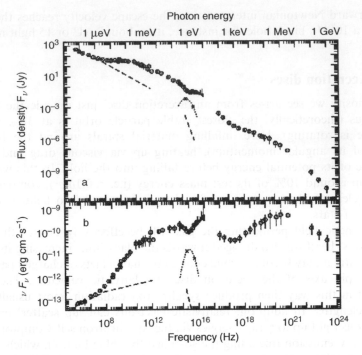

Figure 6.5   The spectrum of 3C273 across all frequency domains (Reproduced with permission from Türler *et al.*, 1999). The top panel shows the flux density F_ν while in the bottom panel νF_ν represents the amount of energy emitted in each spectral range.

electrically so that each component carries the other along with it. Thus the limiting case occurs when L_{nuc} reaches the Eddington luminosity

$$L_{\text{Edd}} = \frac{4\pi G M_{\text{BH}} m_{\text{p}} c}{\sigma_T} \qquad (6.2.4)$$

Numerically this gives

$$L_{\text{Edd}} = 1.3 \times 10^{31} \frac{M_{\text{BH}}}{M_\odot} \text{W} \simeq 3 \times 10^4 \frac{M_{\text{BH}}}{M_\odot} L_\odot \qquad (6.2.5)$$

Although super-Eddington luminosities are possible (for instance, we have implicitly assumed spherical symmetry which is clearly incorrect in detail), realistic luminosities generally cannot exceed the Eddington value by a significant factor. Thus if L_{nuc} is observed to be $\sim 10^{11} L_\odot$, for instance, the central mass must be in excess of $10^6 M_\odot$.

6.2.2   Broad line clouds and the molecular torus

Considering the structure of the inner regions of a Seyfert galaxy in more detail, as we work out from the accretion disc, which extends from an inner edge a few AU from the centre out to perhaps 0.01–0.05 pc, we know from the broad lines, the high ionisation states and the variability timescales that the next component we come to is the BLR. This is composed of individual broad line clouds with densities above 10^{16} atoms per m^3, photoionised by the central source. From the observed Hα emission we can estimate that the total combined mass of the clouds is perhaps $100M_\odot$. Since short-wavelength radiation below the Lyman limit is absorbed by any hydrogen it encounters, the fact that we do see flux below 91.2 nm from the nucleus implies that the clouds cannot completely surround the central engine; the 'covering factor' is less than unity. The strength of the broad emission lines varies if the nuclear output changes, so by correlating changes in total flux with changes in the high ionization lines, the time lag of just days again implies that some BLR clouds must be within a few light days ($\sim$0.01 pc) of the central engine. Low ionisation lines show a longer lag time indicating that they are further out, probably around 1 pc. As we saw above, we also see narrow lines from forbidden transitions. These must arise where the densities are less than about 10^{14} atoms per m^3.

Next comes the molecular torus with an inner edge at a radius of about 1 pc and extending out to 100 pc or more. The NLR occurs in the range 10–100 pc, perhaps extending to 1 kpc. Recall that in Seyfert 2s we do not see the nucleus directly as it is obscured by the dust torus. In other words, all Seyferts are

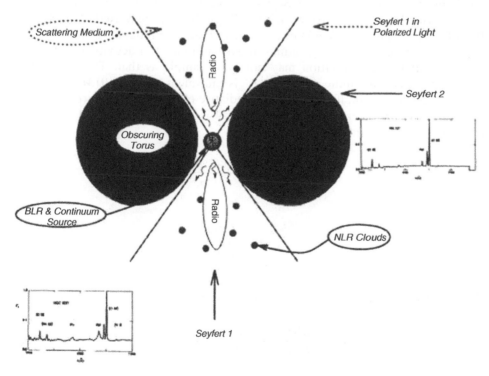

Figure 6.6 Schematic cross-section through the inner few hundred parsecs of a Seyfert galaxy (Reproduced with permission from Brinks and Mundell, 1996).

physically the same and only vary in appearance because of the angle from which we view them. Thus the schematic of a Seyfert illustrated in Figure 6.6 represents *all* the types of Seyfert; we just have to imagine viewing it from different directions, as indicated around the periphery. This is one aspect of 'unification' models of AGN. As noted earlier, the greater number of Seyfert 2 s compared to Seyfert 1 s implies that the 'opening angle' of the torus is quite small and we have to be quite close to the axis to see down to the central engine. Intermediate lines of sight close to the edges of the torus give rise to the gradually changing characteristics of the intermediate Seyfert types. The molecular torus is required to be very dense, with a column density in excess of $10^{25}\,\mathrm{cm}^{-2}$, in order to account for the strongly absorbed X-ray spectrum in Type II systems. The inner edge of the torus is probably also responsible for reflecting some of the X-rays in Type I systems, imprinting the characteristic iron absorption features on the spectrum. Notice also that the torus will block the ionising flux from the central source, so that ionised material or clouds should appear only in two 'ionisation cones', around the axis of the torus on either side of the centre.

6.2.3   Timescales

Nearly all Seyferts are spirals, and about 10% of early type spirals, that is those with significant bulge components, are Seyferts. Notice that this could be due to 10% of spirals having long-lived active nuclei, or all spirals showing Seyfert characteristics 10% of the time. The latter could be as a result of the intermittent availability of fuel, though the presence of large quantities of gas but no nuclear activity in our Galaxy cautions that the real situation may not be as simple as that. The mass inflow requirement is actually quite small; $10^{11}L_\odot$ is equivalent to only $0.01 M_\odot c^2$ per year.

In a little more detail, if we expect the radiative efficiency ϵ to be around 5–10% (i.e. 5–10% of the accreted rest mass energy is radiated away) and we assume radiation at the Eddington rate, then

$$L = \epsilon c^2 \frac{\mathrm{d}M}{\mathrm{d}t} = \frac{4\pi G M m_\mathrm{p} c}{\sigma_T} \tag{6.2.6}$$

or

$$\frac{\mathrm{d}M}{\mathrm{d}t} = \frac{M}{\epsilon t_\mathrm{Edd}} \tag{6.2.7}$$

where we have defined the Eddington time scale by

$$t_\mathrm{Edd} = \frac{Mc^2}{L_\mathrm{Edd}} = \frac{\sigma_T c}{4\pi G m_\mathrm{p}} = 4.4 \times 10^8\,\mathrm{yr} \tag{6.2.8}$$

Notice that this is the same for all accreting black holes and is similar to the dynamical time for a large galaxy.

We can then solve the differential equation (6.2.7) to obtain a timescale for the accretion

$$t_{acc} = \epsilon t_{Edd} \ln(M/M_0) \tag{6.2.9}$$

where M_0 is the original or 'seed' mass of the black hole. Looked at the other way round,

$$M \propto e^{t/\epsilon t_{Edd}} \tag{6.2.10}$$

So the black hole grows by a factor e in a time ϵt_{Edd}. Note that the black hole grows faster if its radiative efficiency is low, since more mass can be accreted without the Eddington luminosity being reached.

The most luminous AGN are of course quasars. These share most of the properties of Seyfert nuclei in terms of their observed spectra, and the basic model for a quasar is essentially the same as that sketched above for Seyferts. However, although it was originally thought that the galaxies underlying radio-quiet quasars would turn out to be spirals, like the hosts of Seyferts, detailed high resolution imaging with the HST has demonstrated that both radio-loud and radio-quiet quasars reside in giant elliptical galaxies.

From the above calculation of timescales, if $\epsilon = 0.05$, a quasar with luminosity $10^{12}L_{\odot}$ requires an accretion rate of $8 \times 10^{22}\,\mathrm{kg\,s^{-1}}$ or about $1.3M_{\odot}\,\mathrm{yr^{-1}}$. Even $10^9 M_{\odot}$ of 'fuel' would suffice only to keep the quasar going for about 700 Myr. Thus we may expect quasar activity to be only a temporary phase in a galaxy's evolution. The known existence of black holes in galaxies with no nuclear activity may point to the whereabouts of once active, but now dormant, galactic nuclei. Indeed, the very tight correlation between spheroid velocity dispersion and black hole mass, where it can be measured (section 3.11), argues that all ellipticals and spiral bulges should contain SMBH. If we estimate the total present-day density in black holes by suitable scaling of the bulge (baryonic) masses – approximately a factor $1/500$ – we obtain $\rho_{\bullet} \simeq 2 \times 10^5 M_{\odot}\,\mathrm{Mpc^{-3}}$, which is consistent (to within a factor 2) with the mass which must have been accreted to power all observed luminous quasars.

6.2.4   LINERs

There are also less energetic nuclei, referred to as LINERs, for Low Ionisation Nuclear Emission Regions, a fairly self-explanatory cognomen. They occur in possibly 25% of early type spirals. They may be powered by lower luminosity versions of Seyfert nuclei or by bursts of star formation around the nucleus or, indeed, by both. In starburst nuclei the available gas will be exhausted on timescales of only millions of years, but during the burst the nucleus can be highly luminous. This can obviously lead to bright blue nuclei (as seen for instance in non-Seyfert Markarian galaxies), though in other cases the starburst can be completely dust enshrouded, as in some ultra-luminous IR galaxies or the galaxies seen at sub-mm wavelengths. Spectroscopically, AGN can be distinguished from

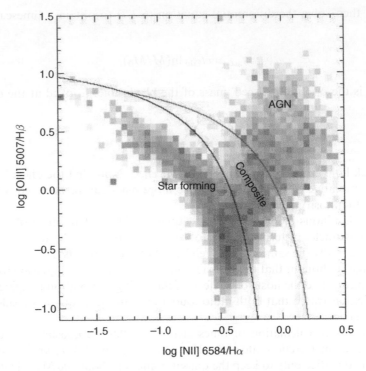

Figure 6.7   The distribution of star-forming galaxies and AGN from the Sloan Survey in the BPT diagram (Reproduced with permission from Brinchmann *et al.*, 2004).

starbursts via their position in the BPT (for Baldwin, Phillips and Terlevich) diagram, a plot of the emission line ratios [OIII]λ5007/Hβ and [NII]λ6584/Hα. Active Galactic Nuclei exhibit relatively large values of each, and so sit in the upper right of the diagram (Figure 6.7).

6.3   Radio galaxies

Another important class of objects, at first sight unrelated to Seyferts and quasars, is that of radio galaxies. In the early days of radio astronomy in the 1940s and 1950s, it was soon found that some of the brightest radio sources on the sky were associated with elliptical galaxies. Indeed, names such as Cen A (previously just plain NGC 5128) derive from these observations – in this case from the brightest radio source in the constellation Centaurus. In many cases the radio sources were double, with two huge Mpc sized lobes either side of the optical galaxy. However, at higher resolution these objects – soon named 'radio galaxies' – also revealed central cores associated with the galaxy nucleus, giving the clue to the connection between radio galaxies and other active galaxies. Many radio galaxies also revealed jet-like structures (as already seen in the optical in M87 by Curtis back in 1918), the jet connecting the central source to the much larger scale lobes. Notice that the largest lobes – they can be up to about 3 Mpc across – must have ages of at least tens of Myr, just to transport the energy over these huge distances.

As we saw in Chapter 4, ordinary spiral galaxies also emit in the radio at some level. However, the name radio galaxy is usually reserved for objects with radio powers above 10^{34} W. The most luminous radio sources can reach 10^{38} W. On the other hand, our Galaxy, as a typical spiral, emits only 10^{30} W in the radio. In common with the spirals, though, the emission mechanism is the synchrotron process, so again we see a spectrum

$$I_\nu \propto \nu^{-\alpha} \tag{6.3.1}$$

with a spectral index $\alpha \simeq 0.7-1.2$. The central cores, on the other hand, have much flatter spectra with $\alpha \simeq 0.4$ or less and are therefore best seen at high frequency, where the lobe emission is much reduced.

Amongst the 'double' radio sources, those with flux densities up to about 10^{25} W Hz^{-1} (at a normalising frequency of 1.4 GHz) have lobes which are brightest in the centre, the ends being fainter ('edge darkening') and showing steeper radio spectra. The jets are double sided and continuous and much brighter than the lobes. These are known as Fanaroff–Riley Class I, or just FRI, galaxies. An example is Cen A. The brighter sources like Cygnus A, which emit upwards of 10^{35} W across the GHz frequency regime, have lobes which are 'edge brightened', have steeper spectra in the inner regions, usually contain much smaller, kpc-sized, 'hot spots' and are classified as FRIIs. Their jets are usually one sided, or at least very asymmetric, and clumpy. There is also less contrast between the jet and lobes, even though the jets themselves are brighter than in FRIs. The 'core' radio source, in what is really a triple, corresponds to the self-absorbed base of the jet. Much of the energy comes from a region only a few parsecs in size, the same scale as the NLR and molecular torus which we discussed earlier in the context of Seyferts.

FRIs appear to be hosted by the most luminous ellipticals and cD galaxies, while FRIIs are associated with normal giant ellipticals. We can again assume that the central engine is a black hole – direct observation of galaxies like M87 reveals the characteristic high velocities of the gas just outside the nucleus, consistent with SMBH of mass $10^9 M_\odot$. Thus extrapolating to the less powerful systems, we can surmise that all $L > L_*$ ellipticals house black holes of a million solar masses or more.

For sources in rich clusters, the interactions between the cluster intergalactic medium (IGM) and the radio plasma can distort the radio structures. Since smaller ellipticals in a cluster have significant peculiar velocities, the radio plasma tends to be swept back, like a wake, as they plough through the intergalactic medium. Depending on how large the effect is, they can appear as either 'wide angle tails' or 'narrow angle tails' (WATs and NATs).

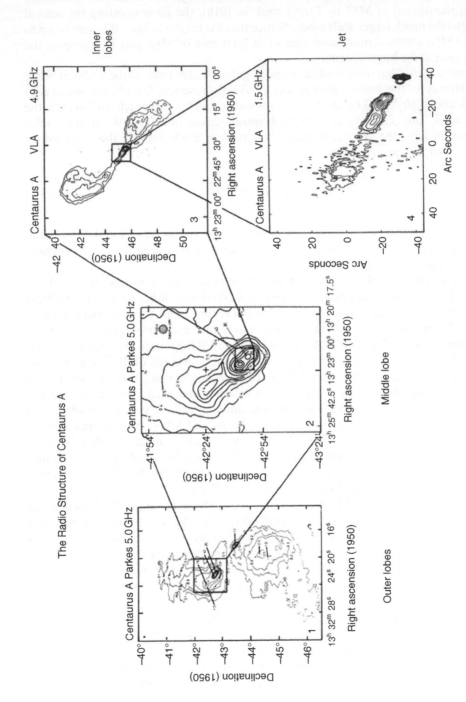

Figure 6.8 A radio image of Cen A, showing the double lobe structures and jets on a variety of scales (Reproduced with permission from Burns, Feigelson and Schreier, 1983).

6.4   Synchrotron emission

The radio lobes are fed from the central core, relativistic plasma being channelled out along the jets. The jet material ploughs into the external medium, creating shocks and dumping energy at what is known as the 'working surface'. The lobes typically have an implied energy content around 10^{53} J (see below) which can be supplied by a 10^{10} $L_\odot$ source in 1 Gyr, for example. As we saw above, some radio galaxies show twin, more or less equal, jets but frequently we see one-sided or asymmetric jets. However, all are thought to have double jets; in reality, the difference in *appearance* being due to relativistic beaming. As discussed in section 4.8, particles moving relativistically at speed $\nu \sim c$ radiate primarily in the forward direction, roughly speaking into a beam of opening angle $1/\gamma$, where γ is the Lorentz factor $(1 - \nu^2/c^2)^{-1/2}$. Thus the approaching jet is brightened and the receding one is dimmed. The synchrotron jets are sometimes also seen in the optical (as in M87) and at X-ray wavelengths.

In order to examine the synchrotron process in more detail, consider first non-relativistic electrons (i.e. with $\nu \ll c$) in a magnetic field B. From Maxwell's equations, the force on an electron will be

$$F = -eB \times v \qquad (6.4.1)$$

where e is the electronic charge. For simplicity, consider the case where v is perpendicular to B, so that we need only consider the amplitude of the vector terms. Remember, though, that the cross-product implies that F is perpendicular to the

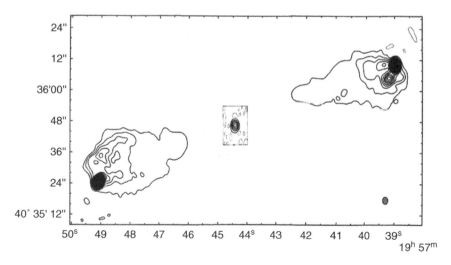

Figure 6.9   The very high power radio galaxy Cyg A was first detected by Reber in 1944. The low resolution radio map shown here (Reproduced with permission from Hargrave and Ryle, 1974) emphasises the hot spots at the ends of the lobes.

plane defined by $\boldsymbol{B}$ and $\boldsymbol{v}$. Thus the electron will move in a circle of radius r_g around a magnetic field line with acceleration

$$\frac{v^2}{r_g} = \frac{eBv}{m_e}$$

(6.4.2)

so

$$r_g = \frac{m_e v}{eB}$$

(6.4.3)

This is known as the Larmor- or gyro-radius. The frequency of the emitted radiation is then the gyro-frequency

$$\nu_g = \frac{\nu}{2\pi r_g} = \frac{eB}{2\pi m_e}$$

(6.4.4)

the frequency with which the electron 'orbits' the field line. The emitted radiation has a dipolar pattern ahead of and behind the source electron.

If, instead, the electron is moving relativistically, with Lorentz factor γ, we can replace the momentum $p = mv$ by the relativistic momentum $p = \gamma mv$ to obtain the orbital frequency $\nu_o = \nu_g/\gamma$. The radiation is beamed into a cone of opening angle $\sim 1/\gamma$ about the instantaneous direction of travel, as before, so as the electron circles the field line the beam will sweep across the line of sight to the observer. The electron will orbit through an angle $1/\gamma$ while the pulse remains visible, which will take it a time $r_g/\nu\gamma$. But it will now be nearer to the observer by the amount r_g/γ (assuming

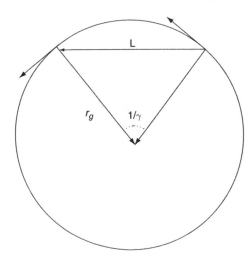

Figure 6.10 Schematic view of the relation between the gyro-radius and pulse width for a relativistic particle.

$\gamma \gg 1$), so the light travel time at the 'end' of the pulse will have been reduced by $r_g/c\gamma$. Thus the width of the pulse (in time) will be

$$\Delta t \simeq \frac{r_g}{\gamma v} - \frac{r_g}{\gamma c} = \frac{r_g}{\gamma v}\left(1 - \frac{v}{c}\right) \simeq \frac{r_g}{\gamma v}\frac{1}{2\gamma^2} \tag{6.4.5}$$

giving a frequency

$$\nu \simeq \frac{1}{\Delta t} \simeq \frac{2\gamma^3 v}{r_g} \sim \gamma^2 v_g \tag{6.4.6}$$

A more complete treatment reveals that there will be a spread in observed frequencies around

$$\nu_s = \frac{3}{2}\gamma^2 v_g \propto \gamma^2 B \tag{6.4.7}$$

Recall that the relativistic particle energy $E = \gamma m_e c^2$, so $\nu_s \propto BE^2$ and the energy of the emitted photons $h\nu_s$ is proportional to the *square* of the electron's energy. Reintroducing the numerical factors, for B in tesla

$$\nu_s = 4.2 \times 10^4 \gamma^2 B \text{ MHz} \tag{6.4.8}$$

or, for energies in eV,

$$\nu_s = 0.11E^2 B \text{ Hz} \tag{6.4.9}$$

Typical galactic fields of 10^{-9} T and electron energies of GeV then imply frequencies of 10^8 Hz, in the radio region.

6.4.1   Energy loss rates

Since the electrons are radiating, they are evidently losing energy. In general, a non-relativistic electron with acceleration a will generate an electric field at distance r (in a direction at an angle θ to the acceleration vector) given by

$$\mathcal{E} = \frac{ea\sin\theta}{4\pi\epsilon_0 c^2 r}$$

(where ϵ_0 is the permittivity of the vacuum) and the energy flux of this field is its Poynting vector

$$S = \frac{\mathcal{E}^2}{\mu_0 c} \tag{6.4.10}$$

Thus, recalling that μ_0 and ϵ_0 are related by $\epsilon_0\mu_0 = 1/c^2$, the energy lost by the electron into a solid angle $\Delta\Omega = 2\pi\sin\theta\,\Delta\theta$ is

$$\Delta\left(\frac{dE}{dt}\right) = -Sr^2\Delta\Omega = \frac{e^2 a^2 \sin^3\theta\Delta\theta}{8\pi\epsilon_0 c^3} \tag{6.4.11}$$

Integrating over all solid angles this becomes

$$\frac{\mathrm{d}E}{\mathrm{d}t} = -\frac{e^2 a^2}{6\pi\epsilon_0 c^3} \qquad (6.4.12)$$

In order to be able to use an equivalent expression for relativistic particles, it is simplest to transform to the electron's rest frame. In this frame the magnetic field (call it B') is moving and induces an electric field $\mathcal{E}'$, but the energy loss rate is the same (a 'Lorentz invariant'). Also, since the velocity in this frame is zero, there is no $v \times B$ force and the acceleration a' will depend only on $\mathcal{E}'$. Specifically

$$\mathcal{E}' = \gamma v B_\perp \qquad (6.4.13)$$

(where $B_\perp$ is the perpendicular component of the magnetic field) and

$$m_e a' = -e\mathcal{E}' \qquad (6.4.14)$$

Thus

$$-\frac{\mathrm{d}E}{\mathrm{d}t} = -\frac{\mathrm{d}E'}{\mathrm{d}t'} = \frac{e^2 (a')^2}{6\pi\epsilon_0 c^3} = \frac{e^4 \gamma^2 v^2 B_\perp^2}{6\pi\epsilon_0 m_e^2 c^3}$$

$$= 2\sigma_T c \gamma^2 \frac{v^2}{c^2} \frac{B^2}{2\mu_0} \sin^2\theta \qquad (6.4.15)$$

where $\sigma_T \equiv e^4/6\pi\epsilon_0^2 c^4 m_e^2$ is the Thomson cross-section. Notice that the energy loss rate is proportional to m^{-2} so protons lose energy much more slowly than electrons.
   For relativistic electrons with $v \simeq c$

$$-\frac{\mathrm{d}E}{\mathrm{d}t} \simeq 2\sigma_T c \gamma^2 u_B \sin^2\theta \qquad (6.4.16)$$

where u_B is the magnetic energy density $B^2/2\mu_0$. Alternatively, feeding in the values of the constants, for B in T and E in eV, this is

$$-\frac{\mathrm{d}E}{\mathrm{d}t} = 3.8 \times 10^{-7} E^2 B^2 \sin^2\theta \text{ eV s}^{-1} \qquad (6.4.17)$$

Averaging over all possible angles (with probability $\propto \sin\theta$) brings in a factor 2/3, so finally

$$-\frac{\mathrm{d}E}{\mathrm{d}t} = \frac{4}{3}\sigma_T c \gamma^2 u_B \qquad (6.4.18)$$

Again inserting numerical values this becomes

$$-\frac{dE}{dt} = 1.0 \times 10^{-14} B^2 \gamma^2 \text{ W} \tag{6.4.19}$$

Notice that high energy electrons lose their energy fastest. All other things being equal, we can just integrate the equivalent expression

$$\frac{dE}{E^2} \propto B^2 dt \tag{6.4.20}$$

to give a timescale for the energy loss via

$$\frac{1}{E} \propto B^2 t \tag{6.4.21}$$

The energy loss time, say to lose half the original energy, will then be

$$t_s \propto \frac{1}{EB^2} \tag{6.4.22}$$

or more precisely

$$t_s = 1.7 \times 10^{-11} B^{-2} \gamma^{-1} \text{ yr} \tag{6.4.23}$$

In terms of the emitted frequency, we can alternatively write this as

$$t_s = 34 \left(\frac{10^{-9} \text{ T}}{B}\right)^{3/2} \left(\frac{1 \text{ GHz}}{\nu_s}\right)^{1/2} \text{Myr} \tag{6.4.24}$$

Thus, for the fields of order 10^{-9} T thought to be typical of the lobes, the lifetime of electrons radiating at GHz frequencies will be tens of Myr. For the electrons producing the much higher frequency X-ray synchrotron emission in jets, where the magnetic field is probably ten times higher, too, the lifetime can be no longer than 100 years. In addition, electrons can lose energy by inverse Compton scattering, at a corresponding rate $(4/3)\,\sigma_T c \gamma^2 u_\gamma$, where u_γ is the energy density in the radiation field. In the process, they increase the energy of radio or optical photons by factors γ^2 to those of X-rays or γ-rays. Indeed, in addition to scattering off the photons produced by the galaxy itself, the electrons may also interact with the 'ambient' photons of the Cosmic Microwave Background.

6.4.2 Acceleration processes

Of course, in a real astrophysical plasma, there will be electrons of a range of energies. If observed cosmic ray electrons in our Galaxy are a reasonable guide, then we can expect the various acceleration processes to give rise to a power law energy spectrum

$$N(E)dE \propto E^{-s}dE \qquad (6.4.25)$$

Now the energy lost by electrons of energy between E and $E + dE$ must be the same as the energy of the photons emitted. Thus the intensity I must be given by

$$I(\nu)d\nu \propto E^{-s}dE\frac{dE}{dt} \qquad (6.4.26)$$

so with

$$\frac{dE}{dt} \propto B^2 E^2 \qquad (6.4.27)$$

we have

$$I(\nu)d\nu \propto E^{2-s}dE \qquad (6.4.28)$$

But we already know that $\nu \propto E^2$ so

$$I(\nu)d\nu \propto \nu^{(1-s)/2}d\nu \qquad (6.4.29)$$

Thus we expect a power law emission spectrum with, in our usual notation,

$$\alpha = \frac{s-1}{2} \qquad (6.4.30)$$

For cosmic rays arriving at the Earth we find $s \simeq 2.6$, so we would expect $\alpha \simeq 0.8$, as is indeed seen for the radio spectrum of spiral galaxies like ours. In fact, most other synchrotron sources, including the radio lobes of interest here, also have similar spectral indices, suggesting that there may be some common acceleration mechanism(s) which provides electrons with a 'standard' energy spectrum with $s \simeq 2.6$.

The most likely candidate is the Fermi process wherein electrons accelerate by effectively bouncing off clouds containing magnetic fields. Due to there being slightly more head-on collisions than following ones, there is a net energy gain for the electrons. However, the energy gain is very slow in the original version of the process; the fractional gain in energy per collision is only of order $(\nu/c)^2$ for clouds moving at speed ν and the mean free path (and hence time) between collisions is very long. Furthermore, the particular observed energy spectrum with $s \sim 2.6$ is not explained. On the other hand, we can obtain a 'first order' (in ν/c) Fermi process if we introduce shocks into the picture. The most important feature here is that gas on either side of the shock is travelling at different speeds – clouds approach the shock more rapidly than they leave it behind. Thus two clouds either side of a shock are effectively moving towards each other and a cosmic ray (which will not notice the presence of the shock) can bounce backwards and forwards between them, suffering a head-on collision every time and thus gaining energy at every shock crossing (instead of having nearly equal numbers of energy-gaining and energy-losing collisions).

Suppose we start with n_0 electrons of energy E_0. Since the energy increase is by a factor $(1 + v/c)$ at each collision, after m scatterings the energy is $E_m = E_0(1 + v/c)^m$. But the probability of a cosmic ray electron escaping at each stage turns out to be (v/c), so the number reaching an energy $E_0(1 + v/c)^m$ is $n_0(1 - v/c)^m$. Hence

$$\frac{\ln(n(>E)/n_0)}{\ln(E/E_0)} = \frac{m \ln(1 - v/c)}{m \ln(1 + v/c)} \simeq -1 \qquad (6.4.31)$$

Thus $n(>E) \propto E^{-1}$, so we must have

$$n(E)\mathrm{d}E \propto E^{-2}\mathrm{d}E \qquad (6.4.32)$$

close to the observed form. We can reach the required $s \simeq 2.6$ by including other energy losses, curved shock fronts or other details.

In practice, the synchrotron spectrum turns down from the pure power law at both ends. At low frequencies, where the source becomes optically thick to its own radiation ('synchrotron self-absorption'), the spectrum switches over to $\alpha = -2.5$. In this case (cf. equation 4.6.8) $I \propto k_B T v^2$ and for relativistic particles $3k_B T = \gamma m_e c^2$, so that (recalling $\gamma^2 \simeq v/v_g \propto v/B$) we obtain

$$I \propto v^{5/2} B^{-1/2} \qquad (6.4.33)$$

At very high frequencies, the fall-off of the spectrum steepens due to the rapid energy losses noted above.

6.4.3   Energy densities

Again, given that the energy loss $\mathrm{d}E/\mathrm{d}t$ can be equated to the emitted radiation, it is evident that the radio luminosity from a source of volume V and electron density n must follow

$$L \propto nVE^2 B^2 \qquad (6.4.34)$$

So the total energy of the electrons is

$$U_e = nVE \propto \frac{L}{EB^2} \propto LB^{-3/2} \qquad (6.4.35)$$

since for electrons radiating at a particular frequency, $E^2 B$ is constant. Similarly the total magnetic energy is

$$U_B \propto B^2 V$$

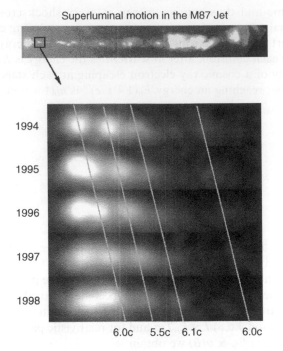

Superluminal motion in the M87 Jet

1994

1995

1996

1997

1998

6.0c   5.5c   6.1c   6.0c

Figure 6.11   HST images of the jet in the nearby radio galaxy M87. The upper panel shows the overall jet (the galaxy is to the far left) while the lower montage shows the apparent superluminal motion of several knots (Reproduced with permission from Biretta, Sparks and Macchetto, 1999).

We can notice that the energy of the electrons decreases as B increases, while the magnetic energy obviously increases as B increases. The total energy in both components is therefore minimised at some intermediate value of the field. Differentiating the formula for the total energy gives two terms, proportional to $LB^{-5/2}$ and to VB. Thus it is evident that the minimum total energy must occur at some $B \propto (L/V)^{2/7}$ where the total derivative goes to zero. In detail, this then gives (for L_ν in W Hz^{-1} and V in m^3)

$$B_{\min} = 1.8\, L_\nu^{2/7}\, V^{-2/7}\, \nu^{1/7}\ \text{T} \tag{6.4.36}$$

and

$$U_{\min} = 3.0 \times 10^6\, L_\nu^{4/7}\, V^{3/7}\, \nu^{2/7}\ \text{J} \tag{6.4.37}$$

This is also close to the 'equipartition' value where $U_e = U_B$. In fact it is easy to check that at the minimum energy $U_B = 3U_e/4$. For the lobes of radio galaxies, the equipartition value is typically the 10^{-9} T (10 μG) used above. This is close to the strength of the field in the Galactic disc near the Sun. The total lobe energies then amount to of order

10^{52} to 10^{54} J. Even though the fields can be stronger, the volumes of the jet and core sources are so much smaller that their total energy content is orders of magnitude smaller. Note that in all this we have ignored the contribution from particles other than electrons, which may or may not be a serious omission in different cases. In addition, though the assumption of equipartition or minimum energy seems reasonable, it is not yet particularly well justified observationally. (Calculations of, for instance, the pressure balance between lobes and the external medium, using a combination of radio and X-ray observations, can give some constraints.)

6.5 Jets and superluminal motion

Detailed observations of jets using very long baseline interferometry (VLBI) show them to contain discrete knots in the inner 50 kpc or so (where the jet is brightest). New knots appear from time to time and are seen to move out along the jet axis before fading. The inferred velocities across the sky are often 'superluminal', that is apparently faster than the speed of light. Suppose a blob of plasma is injected with a relativistic velocity v at an angle θ to the line of sight. After a time Δt in the reference frame of the galaxy, the blob will have moved a transverse distance $v \sin \theta \Delta t$. However, because of its velocity towards the observer it will also be nearer by $v \cos \theta \Delta t$. Thus a signal (photon) leaving the blob at time Δt takes $v \cos \theta \Delta t / c$ less long to arrive at the observer, and the observer therefore records a time interval

$$\Delta t_{\text{obs}} - \Delta t (1 - v \cos \theta / c) \tag{6.5.1}$$

The apparent transverse velocity therefore becomes

$$v_{\text{obs}} = \frac{v \sin \theta}{1 - v \cos \theta / c} \tag{6.5.2}$$

which can be greater than c.

Notice that the most favourable orientation (for the maximum v_{obs}) occurs when $\cos \theta = v/c$, in which case

$$v_{\text{obs}} = \frac{v \left(1 - v^2 / c^2\right)^{1/2}}{1 - v^2 / c^2} = \gamma v$$

If we observe a transverse velocity of, say, $5c$, we can therefore deduce that the Lorentz factor γ must take at least the value 5 (since the real velocity v cannot exceed c). This translates to $v \geq 0.98c$. Recall that we will see a jet with this γ only if we are within an angle $1/\gamma$ (radians), or about $10°$, of the jet axis, which is also the optimum point for seeing superluminal motion.

Since nearly all the radiation is concentrated in the forward direction, from the observer's point of view the source is brighter by a factor $4\gamma^2$ relative to an isotropic source. In addition, because of Special Relativistic time dilation, the photons emitted by the source in some time interval δt are received by the observer over the interval $\gamma(1 - v/c)\delta t \simeq \delta t/2\gamma$ (assuming $\cos\theta \simeq 1$) and each has its frequency and hence energy increased by a factor $[\gamma(1 - v/c)]^{-1} \simeq 2\gamma$. If the source has a spectral index α, the net effect is an increase in the flux (at a particular frequency) by a factor $\sim(2\gamma)^{3+\alpha}$. In fact, if the overall jet flux is made up of contributions from many blobs, the approaching ones have a lifetime reduced by the factor 2γ as well, so the net effect is an increase by only a factor $(2\gamma)^{2+\alpha}$. The receding jet is effectively dimmed by the same factor, so the flux ratio between the jet and counter-jet will be of order $(2\gamma)^{4+2\alpha}$. Since the radio core usually has $\alpha \simeq 0$, for Lorentz factors of 5 to 10 as deduced from the superluminal motions, the flux ratio will be typically 10^4–10^5, accounting for the many apparently one-sided jets.

6.6   Unification

We have already seen some parts of the 'unification models' for AGN. First, the different classes of Seyfert galaxy are really the same beast, but viewed from different

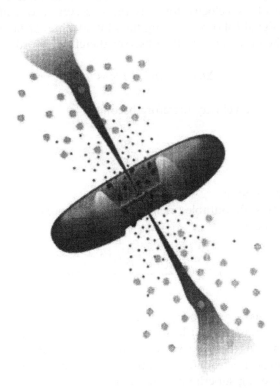

Figure 6.12   The unification scheme for radio loud AGN, showing the molecular torus, emission line clouds and the jets (Reproduced with permission from Urry and Padovani, 1995).

angles. Second, though occurring in elliptical galaxies rather than spirals, in terms of the AGN itself, radio-quiet quasars appear to be simply more luminous versions of Seyferts. They, too, can obviously be seen from different angles and we therefore expect a population of 'buried' quasars. These are almost certainly represented among the (long) unidentified X-ray sources making up the X-ray background, which have now been studied in detail by using the Chandra and XMM-Newton satellites.

Now consider radio AGN. Clearly there are two types of radio galaxies, the FRIs and FRIIs, distinguished by their total radio powers, and probably physically by the nature of their jets. Both types are hosted by giant elliptical galaxies, as are quasars and blazars. If we view an FRI down the jet axis, it is evident that the relativistic boosting will be at its greatest and the approaching jet can completely outshine the relatively low power central engine (i.e. the accretion disc and line emitting regions). In the unification picture, this is the source of BL Lac objects. Allowing (as best as we can, theoretically) for the beaming angles and γ factors, it seems that the relative numbers of FRIs and BL Lacs are consistent with this idea. Both also appear to have their magnetic fields perpendicular to the jet axis, leading to the expectation of shocks as the jet material tries to cross the field lines.

Similarly, we can propose that seeing a more powerful FRII close to the jet axis will give us a radio-loud quasar. In this case we still see line emission as well as the optical continuum from close to the strong central source. As we increase the viewing angle, we might expect to move from OVV quasars with superluminal motion, to quasars with one-sided jets, to unequal doubles, to classical doubles when the jet is in the plane of the sky. The magnetic fields here appear to be parallel to the jets, perhaps suggesting faster jets. Thus we can unify all the radio-loud AGN, as being those AGN which (somehow) produce strong jets and require just two distinct types (of jet) viewed from arbitrary angles to account for all the known types of object.

7 Clusters and clustering

7.1 The distribution of galaxies

In previous chapters, though introducing such ideas as galaxy interactions and the environment dependence of the LF, we have largely treated galaxies as individuals. We will now turn our attention to their distribution in space and the general question of large scale structure in the universe.

The natural post-Copernican assumption is that our Galaxy should not occupy a special position. Averaged on a suitably large scale, the universe should look the same to an observer anywhere. This is enshrined in the 'Cosmological Principle', viz. the universe is spatially homogeneous (the same at all positions) and isotropic (the same in all directions). This is the basis for all the cosmological models of the universe which we shall discuss in the next chapter.

However, it begs the question of what is a suitably large scale. Clearly, it must be large enough to average over a significant number of individual galaxies. Indeed, from what we already know about galaxies occurring in groups and clusters, 'large scale' must be much bigger than a galaxy cluster or, indeed, the distance between large clusters. This clearly implies a scale approaching, or more likely exceeding, 100 Mpc; there are only two rich clusters within 30 Mpc of the Galaxy and the nearest really large cluster, Coma, is about 100 Mpc away. As the Hubble distance c/H_0 is around 4000 Mpc, there is only a rather limited range of scales within the observable universe for which homogeneity can be a reasonable approximation. Nevertheless – and despite suggestions, from time to time, of an infinite hierarchy of systems or of fractal universes – we will assume that homogeneity really does hold

The Structure and Evolution of Galaxies   Steven Phillipps
© 2005 John Wiley & Sons, Ltd

on the largest scales. In this chapter we will consider, *inter alia*, how the observed universe tends towards this uniformity.

In our immediate vicinity, as we have seen, our Galaxy is one of 30 or 40 (most of them rather small) in the LG. This is conventionally assumed to have a radius of 1.5 Mpc, though most of the galaxies actually lie in two small sub-clumps centred on the Galaxy and on M31 (recall Figure 2.14). The LG then runs directly into further structures of similar size and composition, the Sculptor or South Polar Group on one side and the M81 and CVn I[1] groups on the other (cf. Figure 2.15). In general, small groups like these tend to line up into larger filamentary structures and, in our case, stretch around 20 Mpc towards the much larger Virgo Cluster, roughly in the direction of the North Galactic Pole. Virgo has at least 1300 catalogued members, including 150 brighter than $m_B = 14$, or $L \geq 10^9 L_\odot$. The larger environment of the Virgo Cluster, including the surrounding groups, is known as the Local – or more usually these days, Virgo – Supercluster. As pointed out by Holmberg and by de Vaucouleurs, the distribution of nearby galaxies, out to about 30 Mpc, is largely co-planar so they appear to fall mainly in a band around the sky which passes through Virgo (so roughly perpendicular to the Galactic Plane) called the Supergalactic Plane.

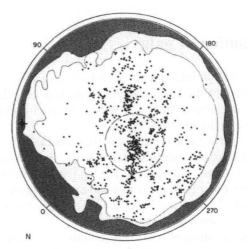

Figure 7.1   The distribution of bright galaxies relative to the Supergalactic Plane, which in this view runs almost vertically through the North Galactic Pole region. The Zone of Avoidance due to dust obscuration in the Galactic Plane is shown shaded around the edges of the map (from de Vaucouleurs, 1976).

7.2   Rich clusters

The first cluster was (inadvertently) discovered by Messier, whose famous list included 11 nebulae in the constellation of Virgo and several more in the neighbouring

---

[1] CVn is the abbreviation for the obscure constellation of Canes Venatici, in whose direction the group's main members appear.

Coma Berenices. William Herschel remarked on this concentration in 1811. Clusters of somewhat fainter nebulae in Coma and Perseus were studied by Max Wolf at Heidelberg in the early years of the 20th century and Wolf also obtained the first useful photographs of a cluster. After the extra-galactic nature of many nebulae was established, many further concentrations were found, especially by Shapley and Hubble. Hubble estimated that 1% of galaxies were found in what he called 'great clusters' of about 500 members.

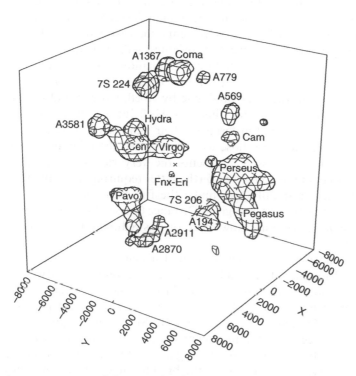

Figure 7.2 A representation of the 3-D distribution of nearby clusters (Reproduced with permission from Hudson, 1993).

Rich clusters are fairly straightforward structures to picture. Basically they are agglomerates of at least a few hundred and often thousands of galaxies in a volume only a few Mpc across. Recall that the galaxy LF tells us that on average there is only one bright galaxy per 200 Mpc^3, so even allowing for the fact that most cluster galaxies will be of fairly low luminosity, it is clear that there is a large density contrast between a cluster and the field. Interestingly, clusters are not much different in size to typical groups, though the latter contain only tens of galaxies. For example, the 'core radius' of Virgo, the radius at which the surface density of galaxies drops to half its central value, is about 1.7° or 0.5 Mpc, while the overall extent of the cluster on the sky is usually judged to be a circle of diameter 12° (≃4 Mpc).

The first really comprehensive cluster catalogues were made in the 1950s. Fritz Zwicky and his colleagues eventually published a seven-volume catalogue of galaxies and clusters of galaxies in the early sixties, but the catalogue with the greatest long-term impact was that produced by George Abell in 1958. His clusters were selected by eye from photographic plates according to a specific set of criteria. To be counted as a cluster in the statistical catalogue of 1682 objects (another 1030 smaller clusters made it into a supplementary list), a structure was required to contain at least 50 galaxies. Furthermore, these had to lie within a circle of radius $1.5\,h^{-1}$ Mpc (about 2 Mpc for our adopted H_0) and to be brighter than $m_3 + 2$, where m_3 is the magnitude of the apparently third brightest galaxy. Notice that the first of these criteria requires knowledge of the distance to the cluster, though the second does not. For those with no prior distance (i.e. redshift) measurements (the vast majority, at that time), Abell utilised the magnitude m_{10} of the 10th brightest galaxy as a standard candle. (This was assumed to be more resistant to contamination from any foreground galaxies than, say, the magnitude of the apparently brightest galaxy.) Equivalent southern clusters were added in the later catalogue by Abell, Corwin and Olowin, making a total of 4073 clusters. The Coma Cluster is Abell 1656, or just A1656, for instance.

Abell Clusters are divided in terms of their number of galaxies into five classes. Richness Class 1 clusters are those which just reach the minimum number require-ment, while Richness Class 5 represents the very richest clusters. Only a small number of clusters are in Class 4 or 5 (the latter requiring more than 300 qualifying members). Clusters just failing to make the full statistical sample are sometimes called Class 0. Clusters were also given Distance Classes 1 to 5, depending on the m_{10} magnitudes, though these are now more or less redundant due to modern redshift measurements.

In terms of morphology, Abell separated his clusters into regular, intermediate and irregular. Though, with much less definite selection criteria, Zwicky's catalogues contained even more, generally less rich, clusters. These were classified as compact, medium compact and open. Other classification systems were produced by Bautz and Morgan (B–M), and Rood and Sastry (R–S). The B–M types run from I to III depending on the degree of domination by the central galaxy, while the R–S types are based on the distribution of the brightest galaxies; for instance, cD clusters have one giant cD galaxy, type L clusters have the brightest few galaxies in a linear structure and type I clusters have them irregularly distributed across the whole cluster. Unsurprisingly, the different schemes correlate to a reasonable extent, dense clusters often being regular, compact, B–M type I and R–S type cD. In Chapters 2 and 3, we noted that elliptical galaxies occur primarily in dense environments, one aspect of the overall morphology–density relation. Thus we should not be surprised that the proportions of ellipticals and spirals (amongst the luminous galaxies) also vary systematically along the same sequence from the regular clusters (around 35% Es, 45% S0s, 20% spirals) to irregular clusters (15%, 35%, 50%).

7.3   Cluster masses

Though galaxies are not as isolated in terms of the ratio of their separations to their sizes, as are stars, we can still usually treat the dynamics of a galaxy cluster in the same way as we did the stellar dynamics in elliptical galaxies in Chapter 3. In particular, we can measure the velocity dispersion. In this case, though, it is determined by the distribution of velocities of individual galaxies, as measured by their Doppler shifts, rather than from the width of the spectral lines in the elliptical.

The measured redshift for a galaxy is composed of two components – one for the overall Hubble flow, that is the cluster's systemic velocity, and one for the peculiar, gravitationally induced velocity of the galaxy inside the cluster. Since the redshift is really a stretching of the spectrum, the overall redshift z is given by the multiplicative relation

$$1 + z = (1 + z_{\rm cl})(1 + z_{\rm pec}) \simeq (1 + z_{\rm cl})\left(1 + v_{\rm pec}/c\right) \qquad (7.3.1)$$

So the peculiar velocity is

$$v_{\rm pec} \simeq \frac{cz - cz_{\rm cl}}{(1 + z_{\rm cl})} \qquad (7.3.2)$$

where $z_{\rm cl}$ is the mean redshift of the cluster galaxies.

For a smallish cluster like Fornax, the velocity dispersion $<v_{\rm pec}^2>^{1/2} = \sigma_r$ is about $350\,{\rm km\,s^{-1}}$, reaching around $700\,{\rm km\,s^{-1}}$ for ones like Virgo. For the largest clusters, $\sigma_r \gtrsim 1000\,{\rm km\,s^{-1}}$. Remember that, as we can measure only radial velocities, this is a 1-D velocity dispersion and in an isotropic cluster the 3-D version will be $\sqrt{3}$ times larger than σ_r.

The first thing we should consider for any possible cluster of galaxies is whether it is likely to be a stable system. In the years after Zwicky's discovery of the large velocities of cluster galaxies (p. 221), it was often postulated that clusters were in fact expanding. We can calculate the time a galaxy will take to cross from one side of a cluster of radius R to the other as simply

$$t_{\rm cr} = \frac{2R}{\sigma_r} \qquad (7.3.3)$$

which gives numerically a crossing time

$$t_{\rm cr} \simeq 6 \times 10^8 \left(\frac{R}{1\,{\rm Mpc}}\right) \left(\frac{\sigma_r}{1000\,{\rm km\,s^{-1}}}\right)^{-1} {\rm yr} \qquad (7.3.4)$$

For typical values of R and σ_r this is of order 1 Gyr, so a galaxy could have travelled across the cluster many times in the age of the universe. If clusters really were unbound, then they would dissipate in only a few hundred Myr.

7.3.1   Virial masses

So, assuming now that the clusters have reached at least approximate equilibrium, we can use the usual virial theorem (equation 3.8.12). Thus, exactly as for elliptical galaxies (equations 3.8.13 to 3.8.16), provided that the positions (r_i) and speeds (v_i) are uncorrelated with the masses of the individual galaxies (m_i), we can write

$$2K = \sum_i m_i v_i^2 = 3\sigma_r^2 M_{cl} \tag{7.3.5}$$

and

$$-U = \sum_{i>j} \frac{G m_i m_j}{|r_j - r_i|} = \frac{G M_{cl}^2}{R_{ef}} \tag{7.3.6}$$

where M_{cl} is the total cluster mass (the sum of the m_i) and R_{ef} represents the effective radius of the cluster (a harmonic mean of the $|r_j - r_i|$). This then gives the cluster's virial mass as

$$M_{cl} = \frac{3 R_{ef} \sigma_r^2}{G} \tag{7.3.7}$$

The value of R_{ef} can be calculated simply by counting galaxies as a function of position, through a neat trick devised by Martin Schwarzschild in the 1950s. Suppose we divide our image of a cluster into vertical strips of width Δx and that there are $S(x)\Delta x$ galaxies in the strip at distance x from the centre (once we have subtracted off the expected number of unassociated background or foreground galaxies, by considering a similar patch of sky away from the cluster). If we can assume a spherical cluster, then S is related to the radial number density distribution $n(r)$ through

$$S(x) = \int_0^{y(R)} 2\pi n\left(\left(x^2 + y^2\right)^{1/2}\right) y \, dy = \int_x^R 2\pi n(r) \, r \, dr \tag{7.3.8}$$

where $y^2 = r^2 - x^2$ and R is the overall radius of the cluster. Taking m as the mass of a typical galaxy, the mass of the cluster is simply

$$M_{cl} = 2m \int_0^R S(x) \, dx \tag{7.3.9}$$

and the potential energy

$$|U| = G m^2 \int_0^R \frac{4\pi r^2 n(r) \, dr}{r} \int_0^r 4\pi r'^2 n(r') \, dr' \tag{7.3.10}$$

But from equation (7.3.8)

$$\frac{dS}{dx} = -2\pi x \, n(x) \tag{7.3.11}$$

which we can substitute into the potential energy equation to obtain

$$|U| = 4Gm^2 \int_0^R \frac{dS}{dx} dx \int_0^x \frac{dS}{dx'} x' dx' \tag{7.3.12}$$

Integration by parts twice then gives

$$|U| = -4Gm^2 \int_0^R S(x) \frac{dS}{dx} x \, dx = 2Gm^2 \int_0^R S(x)^2 dx \tag{7.3.13}$$

Thus, finally we obtain the simple result

$$R_{ef} = \frac{GM_{cl}^2}{|U|} = \frac{2\left(\int_0^R S(x)dx\right)^2}{\int_0^R S(x)^2 dx} \tag{7.3.14}$$

which we can easily evaluate numerically as sums over the original counts in strips. In suitable units, our virial theorem result gives

$$M_{cl} \simeq 7 \times 10^{14} \left(\frac{R_{ef}}{1 \, \text{Mpc}}\right) \frac{\sigma_r^2}{(1000 \, \text{km s}^{-1})^2} \, M_\odot \tag{7.3.15}$$

and thus typical values around 1 to $2 \times 10^{15} M_\odot$ for a rich cluster like Coma and a few $\times 10^{14} M_\odot$ for ones like Virgo. However, if we add up the luminosities of all the galaxies in Virgo, we find a total of $1.3 \times 10^{12} L_\odot$, and for Coma about $6 \times 10^{12} L_\odot$. This implies an M/L ratio of approximately 300 in solar units, far larger than that of individual galaxies, even if we include their higher M/L outer parts, and very much greater than that for any reasonable stellar population. This is the discrepancy first noted by Zwicky in the 1930s and the origin of the first missing or dark mass problem. Note that our derivation of the virial equation is still valid, as in the dark matter case we can include all mass elements, rather than just galaxies, in the summations.

As regards mass densities, from observations of Virgo, for example, we find that the luminosity density in the main body of the cluster is around $4 \times 10^{10} L_\odot \, \text{Mpc}^{-3}$, about 300 times the universal average calculated from the field galaxy LF (Chapter 2). The averaged out surface brightness at the centre is still only $0.5 L_\odot \, \text{pc}^{-2}$, though, more than two orders of magnitude less than at the centre of a typical galaxy disc. We can easily translate these figures into the corresponding stellar mass or overall galaxy densities, simply by multiplying by appropriate M/L ratios.

7.3.2 Gravitational lensing

An alternative route to cluster masses is via gravitational lensing. We already met this idea in section 4.16 with regard to lensing by MACHOs in the Galaxy's halo. However, we must now consider the case where the lens, and probably the source, are extended objects, rather than the point masses we assumed previously. The light rays from any point in a background galaxy which pass close to a large mass like a cluster will be deflected and hence appear to come from a different position on the sky, further from the lens, than they would in the absence of the cluster. Thus each point in the image will be shifted radially outwards and the whole image becomes elongated into an arc, tangential to the direction to the cluster centre. These arcs (and smaller versions called arclets) – first observed in the 1980s and now routinely visible around distant clusters thanks to the high resolution imaging capability of HST – are the key signature of gravitational lensing by clusters. Also, since it is possible for the light from a single source to be bent around either side of the cluster centre (from our point of view), we may see multiple images of the same background galaxy or quasar.

Now, from GR, a light ray passing within a distance b of a mass M is bent by an angle

$$\alpha = \frac{4GM}{bc^2} \tag{7.3.16}$$

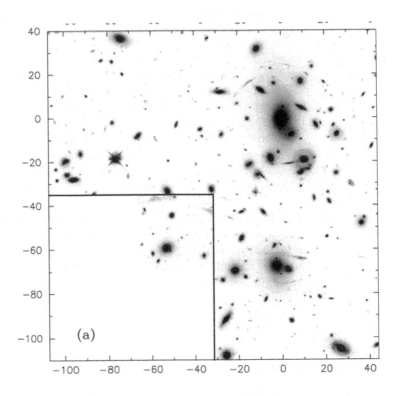

Figure 7.3 Numerous arcs are easily visible in this HST image of the cluster A2218 (Reproduced with permission from Kneib *et al.*, 1996).

For an extended lens, the mass of interest is the mass between the light ray and the cluster centre, and a detailed calculation shows that it is strictly the integrated mass along a column of radius b, call it $M(<b)$, which counts. An isothermal sphere model (as used earlier for elliptical galaxies) with velocity dispersion σ gives a particularly simple result, as the mass is then proportional to b and therefore the deflection angle is independent of b, viz.

$$\alpha = 4\pi \frac{\sigma^2}{c^2} \qquad (7.3.17)$$

For a cluster with $\sigma \sim 1000\,\mathrm{km\,s^{-1}}$ this gives a bending angle $\sim 30''$.

However, the bending angle is not what we directly observe as the shift of position on the sky (Figure 7.4). Without worrying here about the complexities of distance measurements in a curved universe (discussed in Chapter 8), we can say that the angular shift in the position $\theta - \theta_0$ must be a multiple of the bending angle which depends on the relative distances of the lens to the source, D_{LS}, and the observer to the source, D_{OS}, i.e.

$$(\theta - \theta_0)D_{OS} = \alpha D_{LS} \qquad (7.3.18)$$

In addition, if the observer to lens distance is D_{OL}

$$b = \theta D_{OL} \qquad (7.3.19)$$

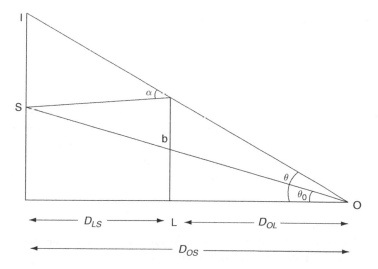

Figure 7.4   The geometry of gravitational lensing. The source at position S appears to be in direction I because of the bending α.

We can combine these to give

$$\theta - \theta_0 = \frac{4GM(<b)\theta}{b^2c^2} \frac{D_{LS}D_{OL}}{D_{OS}} \tag{7.3.20}$$

This can then be written as

$$\theta = \theta_0 \left(1 - \frac{\Sigma(<b)}{\Sigma_{\text{crit}}}\right)^{-1} \tag{7.3.21}$$

where

$$\Sigma(<b) = \frac{M(<b)}{\pi b^2} \tag{7.3.22}$$

is the average (projected) mass density inside b and Σ_{crit} is the critical surface density, defined by

$$\Sigma_{\text{crit}} = \frac{c^2}{4\pi G} \frac{D_{OS}}{D_{OL}D_{LS}} \tag{7.3.23}$$

Notice that provided the density close to the centre of the cluster exceeds Σ_{crit} then a source directly behind the cluster centre will be imaged into a circle – called an Einstein ring – of radius

$$\theta_E = \frac{b_E}{D_{OL}} \tag{7.3.24}$$

where b_E is the impact parameter at which the mean surface density reaches the critical value, i.e.

$$\frac{M(<b_E)}{\pi b_E^2} = \Sigma_{\text{crit}} \tag{7.3.25}$$

Marginally off-centred sources will produce very long thin arcs at essentially the same radius and a central surface density greater than Σ_{crit} is also the criterion for the production of multiple images, as a sufficiently large bending angle is required in order to bring a ray around the 'wrong' side of the cluster centre. In physical units, for a very distant source (say a quasar), so that $D_{OS} \simeq D_{LS}$, the critical density is

$$\Sigma_{\text{crit}} \simeq 2 \times 10^6 D_{OL}^{-1} M_\odot \, \text{pc}^{-2} \tag{7.3.26}$$

for D_{OL} in Mpc. Using this in the earlier relation we obtain a projected mass inside an Einstein ring of angular radius θ_E arcseconds of

$$M(<\theta_E) \simeq 10^8 D_{OL} \theta_E^2 M_\odot \tag{7.3.27}$$

As averaged-out central cluster surface densities are generally of order $100 M_\odot \, \mathrm{pc}^{-2}$, we would not expect to see giant arcs around clusters unless they are at very large distances. However, in most cases there is a large galaxy at the cluster centre, often the dominant cD, so the actual central density is likely to be that appropriate to such a galaxy, at least $10^4 M_\odot \, \mathrm{pc}^{-2}$ and giant arcs and multiple images are actually common phenomena in clusters at distances of a few hundred Mpc. Notice that giant arcs with radii of tens of arcseconds in clusters at distances of hundreds of Mpc imply masses in the central 100 kpc of order $10^{14} M_\odot$.

In general, the close-in giant arcs and multiple images are therefore most relevant to determining the mass of the innermost parts of the cluster. (We can determine the mass of individual, non-cluster, galaxies in this way, too.) In order to obtain the mass of the cluster as a whole, it is necessary to consider the smaller arcs. In this case, it is apparent that if we start with a circular background galaxy, then the axis ratio of the resulting lensed image will depend on how far outwards the image is shifted. Taking the isothermal sphere case, for simplicity, there is no change in the dimension of the image in the radial direction (the direction to the cluster centre). But, in terms of our earlier parameters, the tangential stretching of the image must be by a factor θ/θ_0 which is evidently just

$$A = \frac{\theta}{\theta - \alpha D_{LS}/D_{OS}} \tag{7.3.28}$$

Thus measuring A (which is also the amplification in the brightness of the image, since surface brightness is conserved in lensing) allows us to obtain α and hence the cluster mass. Of course, we do not know that any individual galaxy image would have been circular in the absence of the lens, so we must use statistical properties of the image shapes to determine the lensing potential due to the mass distribution.

7.3.3   X-ray clusters

Nowadays, clusters are known to contain intergalactic gas as well as the galaxies themselves. This intergalactic (also called intracluster) medium is extremely hot, between 3×10^7 and 10^8 K, and therefore revealed by X-ray observations. The mass of gas involved can then be estimated in the same way as for individual elliptical galaxies (section 3.10), since the X-ray luminosity depends on (the square of) the number of free electrons (essentially the same as the number of protons) in the gas. Both Virgo and Coma contain a few $\times 10^{13} M_\odot$ of X-ray-emitting gas. This is several times the mass in stars but still nowhere near the total mass obtained from the dynamics. Indeed, from studies of numerous clusters, it is found that the sum of the stellar masses of the galaxies plus the mass of the gas is usually around 15% of the dynamical mass.

As a check on the overall cluster mass, we can also calculate the mass, or equivalently the depth of the potential well, which is required to hold in the hot gas. Exactly as for the elliptical galaxy case, assuming hydrostatic equilibrium and equal numbers of electrons and protons, this will be

$$M_{\mathrm{cl}}(r) = \frac{2k_B}{m_p G} \frac{r^2}{\rho(r)} \frac{\mathrm{d}}{\mathrm{d}r} (-\rho T) \tag{7.3.29}$$

with ρ and T determined from the X-ray surface intensity and X-ray spectrum, respectively. Such measurements are generally consistent with the direct dynamical ones, which is essentially the same as saying that the velocity dispersions for clusters correspond to the same temperatures as actually measured for the gas, i.e.

$$\sigma_r^2 \simeq \frac{k_B T_X}{m_p} \tag{7.3.30}$$

For instance, for Coma $k_B T_X \simeq 8.5\,\mathrm{keV}$ corresponds quite closely to the observed $\sigma_r \simeq 1100\,\mathrm{km\,s^{-2}}$. The distribution of the X-ray gas is not exactly that expected in the simplest case of an isothermal sphere, though, and is often described by a more general 'beta model' which has a surface brightness profile

$$\Sigma_X(R) = \Sigma_X(0)\bigl(1 + R^2/R_c^2\bigr)^{(-3\beta+1)/2} \tag{7.3.31}$$

where β effectively measures the ratio between the X-ray temperature and the velocity dispersion. (Note that this is the projection of a similar density profile but with exponent $-3\beta/2$.)

Iron lines in the X-ray spectra show that the metallicity of the intracluster gas is perhaps surprisingly high, at typically one-third of the solar value. Given the mass ratio of the stars to gas, this implies that the majority of a cluster's heavy elements are in the intergalactic medium, not the galaxies. Assuming that the processing of primordial gas had to take place in stars in galaxies, this in turn implies a very large loss of gas from galaxies via galactic winds or via 'ram pressure stripping' as the galaxy ploughs through the dense intergalactic medium. (Recall the HI deficiency in cluster spirals mentioned in section 4.12.) Even so, we would also expect a significant fraction of the gas to have been there since the cluster formed, unless the galaxy formation process was extremely efficient and used up all the available gas, which seems rather unlikely. The galactic winds would then presumably 'stir' the metal-enriched processed gas into the primordial metal-free gas.

In many clusters, the density of gas at the centre is so high that cooling processes should be very efficient and the gas would cool out of the X-ray emitting regime in much less than a Hubble time. Unless reheated in some way, it should therefore sink to the bottom of the potential well as a 'cooling flow'. The fate of this cooling gas remained mysterious for many years, as it appeared that tens or hundreds of solar masses per year should be condensing (into something invisible) at the centres of such clusters. More recently, better observations have reduced the implied cooling masses and it has been realised that radio sources may provide a suitable reheating mechanism.

7.3.4   X-ray luminosities

We should also expect the overall X-ray luminosity of a cluster L_X to correlate with its mean temperature, since each reflects the mass. In a simple case, we would have

L_X proportional to the volume (r^3) filled by the X-ray emitting gas-times its emissivity. The latter depends on the particle density (n) squared and the cooling function, which we saw in section 3.7, is typically $\Lambda(T) \propto T^{1/2}$. Thus

$$L_X \propto r^3 n^2 T^{1/2} \tag{7.3.32}$$

If f_x is the fraction of the total mass which is in hot gas, then this is also

$$L_X \propto f_x^2 \rho^2 r^3 T^{1/2} \tag{7.3.33}$$

Now $T \propto \sigma^2 \propto M_{cl}/r \propto \rho r^2$, so $r^3 \propto T^{3/2} \rho^{-3/2}$ and we end up with

$$L_X \propto f_x^2 \rho^{1/2} T^2 \tag{7.3.34}$$

Thus if all clusters have similar densities and hot gas fractions (i.e. are just scaled up or 'self similar' versions of one another, then

$$L_X \propto T^2 \tag{7.3.35}$$

Observation shows that this is only approximately true for high mass hotter clusters ($L_X \sim T^3$) and is a very poor fit for cooler clusters ($L_X \sim T^5$). This means that low temperature (mass) clusters are under-luminous. A plausible suggestion to solve this problem is the so-called 'entropy floor', some preheating mechanism having raised the entropy, in this context defined just as

$$S = \frac{P}{\rho^{5/3}} \propto \frac{T}{n^{2/3}} \tag{7.3.36}$$

rather than the proper thermodynamic meaning. In the similarity solution above, $S \propto T$ but too much entropy is observed for small clusters. This will prevent ρ reaching the same value as in larger clusters and keep L_X down.

7.4   Cluster searches

The X-ray emission also provides a way of detecting clusters in the first place. By the time we reach moderately high redshifts (i.e. large distances) standard optical search methods become compromised. As we shall see in the next chapter, the number of observable galaxies increases rapidly at faint magnitudes, so if we need to look to faint limits in order to see the cluster galaxies then there will be a large degree of contamination from non-cluster galaxies. Thus the surface density contrast of the cluster against the general field will be low. X-ray observations overcome this as we see strong extended X-ray emission only from cluster sources. However, at present the X-ray clusters are also limited to moderate distances due to the observational flux limits.

A somewhat related method is to use what is called the Sunyaev–Zeldovich (SZ) Effect. This arises because the X-ray emitting plasma will also scatter photons from the CMB to higher energies. This effectively removes photons from the low frequency Rayleigh–Jeans side of the blackbody curve which would otherwise arrive at our detectors from the direction of the cluster, so that we see a microwave decrement. Although relatively few clusters have yet been observed in this way, next generation instruments promise all-sky surveys of the CMB which will produce large samples of clusters detected via this effect, which is equally applicable at all redshifts.

We may note, in passing, that the SZ Effect also provides a measure of distances entirely independent of the distance scale based on individual galaxies. Since the X-ray emission at any point is dependent on the square of the particle density times the path length through the cloud, while the CMB decrement just depends on the number of scattering particles along the line of sight, that is the electron density times the path length, we can disentangle the two terms. Thus we can obtain a physical measure of the 'depth' of the cluster. Assuming spherical symmetry, this should match the extent on the sky, so we can use the measured angular size of the cluster to deduce its distance.

In fact, we can surmount the detection problems in the optical, to a degree, by using colour as a second selection criterion. We know from section 3.5 that giant ellipticals all have rather similar red colours. Therefore, if we search for clumps of galaxies which are both of similar magnitude and of similar colour (or better still follow the colour–magnitude relation for ellipticals), then we can be confident that we are seeing a genuine cluster. We should, of course, allow for the fact that galaxy k-corrections make higher redshift elliptical galaxies look even redder than local ones. Such searches for the 'red sequence' are successful in pushing cluster detection out to redshifts around $z = 1$.

The other traditional way of finding clusters is to look *around* objects of known redshift. Since radio galaxies are luminous ellipticals and ellipticals are often in rich clusters, we can use bright radio sources as potential markers. For many years, the most distant known galaxy and cluster was that associated with 3C295 – Minkowski's cluster, at $z = 0.46$, identified in 1960. As above, we may in particular search for associated red galaxies, or we may look for star-forming galaxies with strong emission lines at the same redshift as the radio galaxy. Since quasars also inhabit elliptical galaxies, one might seek to use them to find even higher redshift clusters. However, in practice, at the very large look-back times explored in this way, the environment of the quasars may often be less populous than that of present day giant ellipticals. Indeed, quasars often seem to occur on the peripheries of clusters. Nevertheless, if rich, massive clusters can be found at high redshift then this imposes interesting constraints on the structure formation process, due to the relatively short time available for its development.

In fact, we can see in very general terms that clusters should be of fairly recent construction. Clearly a structure cannot 'condense' out of the overall expanding universe unless its density exceeds the mean. Writing the present day mean density of the universe as ρ_0, galaxies today have density contrasts $\rho/\rho_0 \sim 10^6$, so in principle can have formed when the universe was very much smaller and denser (and therefore younger) than it is now. Clusters, on the other hand, have $\rho/\rho_0 \sim 10^2$ so could only form when the universe was at least $\sim 1/5$ of its present size (and a more accurate calculation gives a

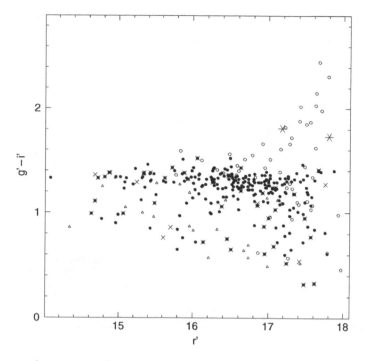

Figure 7.5 A colour–magnitude plot for the region of sky containing the cluster A2255, showing the 'red sequence' of early type galaxies in the cluster (from Miller and Owen, 2003).

value even closer to 1). Superclusters should be even more recent structures, effectively forming only now; the Virgo SC, out as far as the LG, has $\rho/\rho_0 \simeq 3$.

In some cases we can observe the continuing formation of clusters directly, as we see neighbouring structures (other smaller clusters or groups) falling in and merging with the main concentration. This is particularly well seen in X-ray observations of sub-structure (Figure 7.6). These considerations also imply that the distribution of galaxies is evolving with time, becoming clumpier.

7.5   Galaxy groups

As we noted back in section 2.10, galaxy groups are the commonest environment for galaxies. Indeed as long ago as 1938, Zwicky was led to propose that *all* galaxies are in groups or clusters of some size, and this seems to be borne out by modern observations. Isolated, lone galaxies are very rare. The LG is typical of the smaller groups (~3 giant galaxies and ~30 dwarfs), and there is then a continuum all the way up to the smallest clusters (the distinction at the interface being semantic rather than physical).

Crossing times for groups can be long, of order a Hubble time, since they have similar dimensions to clusters (a few Mpc), but much lower galaxy velocities ($\sim 100 \, \mathrm{km \, s^{-1}}$). Nevertheless, if we do assume them to be in equilibrium and apply

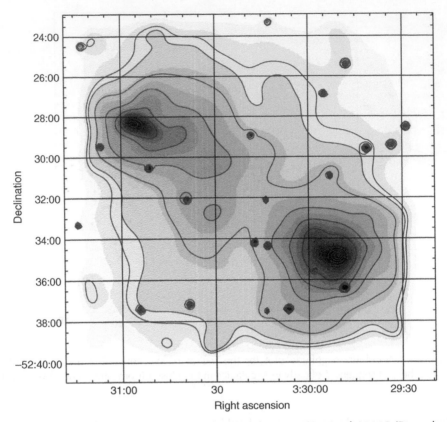

Figure 7.6   A Chandra X-ray image of the merging clusters A3128 and A3125 (Reproduced with permission from Rose *et al.*, 2002).

the virial theorem then we find $M/L \sim 150$–200, again not dissimilar to clusters. Groups must therefore contain substantial amounts of dark matter, and they also follow clusters in containing X-ray emitting gas in many cases. The lower velocity dispersion would lead us to expect lower gas temperatures ($\sim 10^7$ K, or energies ~ 1 keV), and a temperature–X-ray luminosity relation as discussed in the previous section is indeed observed all the way down from large clusters to small groups. This gas could only be retained if $M/L > 100$, supporting the virial mass estimates.

In the case of the LG, we can make a more or less direct estimate of the group mass if we assume it to be concentrated in and around the two largest galaxies, M31 and the Galaxy. This uses what is known as the 'timing argument' first put forward by Franz Kahn and Leon Woltjer in 1959. We currently see M31, at a distance of 770 kpc, *approaching* the Galaxy[2] at about 125 km s^{-1}. If the two were originally

---

[2] Note that the velocity of importance here is that relative to the centre of our Galaxy, not the heliocentric velocity (i.e. that relative to the Sun) which is usually quoted.

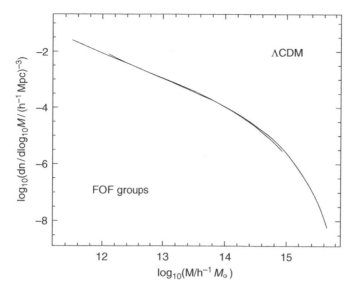

Figure 7.7 The halo mass function derived from N-body simulations of dark matter dominated universes (Reproduced with permission from Jenkins *et al.*, 2001). 'FOF' stands for 'friends of friends', the algorithm used for linking particles into a particular halo.

close together but separating with the Hubble flow, then over the age of the universe, their mutual gravity must have been sufficient to reverse their outward velocity. This is a standard type of two-body dynamical problem and can be solved, with the given boundary conditions, to give the mass required to generate the current infall velocity. This suggests $M_{LG} \simeq 3-4 \times 10^{12} M_\odot$ for a time of 10^{10} years. M31 and the Galaxy will have their next close approach in about 5 Gyr.

In general, in the cold dark matter picture, group galaxies are expected to orbit within common halos of masses $10^{12}-10^{14} M_\odot$, an important number then being the number of separate sub-halos – whether associated with a visible galaxy or not – within the group-sized halo. Theoretical models suggest that for each giant galaxy there should be a very large number of low mass sub-halos, which might be expected to house dwarf galaxies. However the expected mass function (near M^{-2}) is much steeper than the observed luminosity function of, for instance, the LG (L^{-1} or so), so far too few dwarfs are actually seen. On the other hand, the numbers of *systems* as a function of group or cluster mass does seem reasonably in accord with theory.

Finally, we should mention compact groups. These are groups where the projected intergalactic separations are not much greater than the sizes of the galaxies. Unless we are seeing a prolate group end on or a chance temporary alignment of some of the objects in a larger structure, then the compact groups should be particularly conducive to the sorts of interactions discussed in section 3.9 and in a short time dynamical friction should cause the galaxies to merge into one very massive object. However, a suitable population of resulting isolated giant galaxies has not been

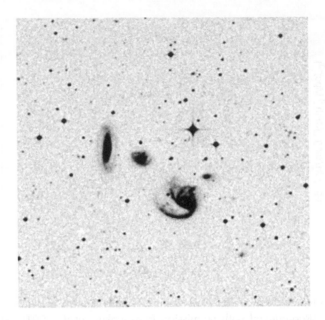

Figure 7.8   The compact group Hickson 63 (UKST image, copyright AAO).

identified, though there have been some claims that these exist in the form of what have been dubbed 'OLEGs' for (X-ray) over luminous elliptical galaxies.

7.6   Intergalactic matter

Clusters and groups evidently contain considerable mass outside of the galaxies. We have already considered the hot X-ray emitting gas and the dark mass. However, clusters also contain other intergalactic material. This is often associated with tidal interactions, as in section 3.9. In groups, individual encounters are expected to be the most important mechanism, as the relative velocities of the galaxies are similar to the internal velocities in the galaxy, the most efficient case for tidal disruption. In clusters, a whole range of related processes have been suggested, including galaxy 'harassment', where numerous high speed encounters as a galaxy flies through the dense environment gradually whittle away its least bound stars.

The cool galactic gas may be removed from its host by ram pressure stripping, as the galaxy travels through a dense existing intracluster medium or may be dragged out by tidal forces. The latter may be responsible for the existence of gas clouds with no obvious optical counterpart, such as the HI cloud in the Leo Group. Gas in tidal tails swept out of galaxies in close interactions will mostly disperse, rather than fall back in, adding to the general ISM. In addition, loosely bound stars will also be dragged into the tidal tails, and then long 'trails' through the cluster and eventually form a population of intergalactic stars. A number of such objects have been discovered in nearby clusters in recent years. Individual 'free flying' red giants have

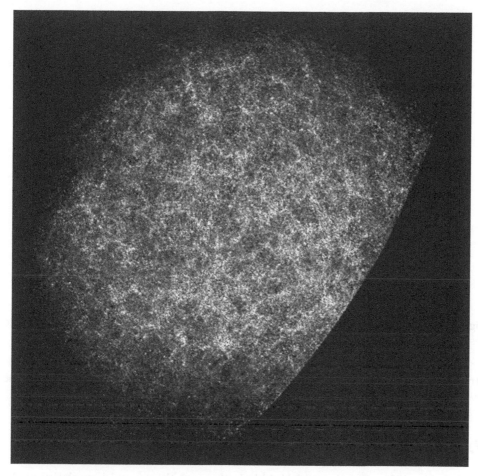

Figure 7.9 The Lick map of the distribution of galaxies over the northern sky (Reproduced with permission from Seldner, Siebers, Groth and Peebles, 1977).

been found in deep HST images of the Virgo Cluster, while narrow band imaging has revealed planetary nebulae (which have very strong line emission at certain wavelengths) not associated with galaxies. Globular clusters are also expected to be removed from galaxies as they pass through the central regions of clusters, creating a general cluster population or adding to the existing population of the central galaxy.

The Centaurus and Coma Clusters show huge arc-like features $\sim$100 kpc long and $\sim$2 kpc wide at a surface brightness level $\sim$27 Bμ($1L_\odot$ pc^{-2}) or less, consistent with tidal debris from a single totally disrupted spiral galaxy. The debris from older events will, presumably, now be spread even more widely through clusters and will generate a general intracluster light. This is very hard to detect as it represents a very small increase over the general sky brightness (perhaps 1%) and has therefore had a somewhat chequered career in terms of claimed detections, with some suggestions that there is more light in the diffuse component than associated with galaxies. Finer scale

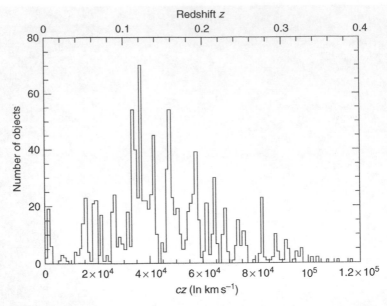

Figure 7.10 The distribution of redshifts of galaxies in the direction of (but mostly much more distant than) the Fornax Cluster (Reproduced with permission from Drinkwater *et al.*, 2000).

structures are easier to detect as we are then looking for small-scale changes in the overall sky brightness.

7.7   Large-scale structure

Back in the 1930s, Shapley and Hubble both noted the general irregularity of the distribution of galaxies on the sky, even outside the obvious individual clusters. Later, maps of the galaxy distribution, such as those derived from the Lick Astrographic Survey (Figure 7.9), or the more recent APM survey using automatically scanned photographic plates,[3] confirmed this non-uniform distribution and suggested a rather filamentary network. This then ties in with what we have already seen of the distribution of other galaxy groups beyond the LG and extending towards the Virgo Cluster.

However, to get a true 3-D picture of the distribution on large scales, we need distances to the galaxies and hence redshift surveys like those discussed in section 2.7. In the 1980s a number of 'pencil beam' surveys of small areas, but extending to quite faint magnitudes, showed that along any given line of sight the redshift distribution is very 'spiky' and suggested the presence of large 'voids', regions empty or nearly empty of galaxies. When combined with results from large area, shallower surveys,

[3] APM is the Automated Plate Measuring machine at the Institute of Astronomy in Cambridge.

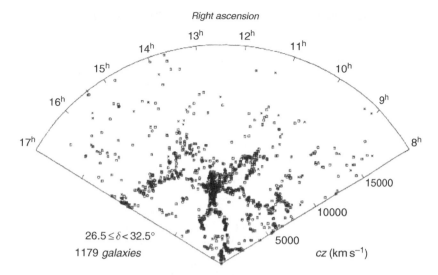

Figure 7.11   The 'stickman' figure seen in the plot of position across the sky (the RA scale in hours around the outside of the wedge) against distance (the redshift scale, in km s^{-1}, at the side of the wedge) for CfA Survey data (Reproduced with permission from Thorstenson *et al.*, 1986).

particularly that carried out at the Harvard-Smithsonian Center for Astrophysics, known as the CfA Survey, it became clear that the 3-D distribution was made up of an interconnecting network of filamentary or planar structures enclosing the empty spaces and making up a foam-like overall topology. This view had been suggested earlier by Jaan Einasto and his colleagues in Estonia[4] and we should also note that the remarkably prescient (and famously irascible) Fritz Zwicky had come up with the idea that the universe was divided up into cells of size 100 Mpc back in the 1930s.

The famous 'stickman' visible in one 'slice' of the CfA Survey[5] is made up of a number of linear features centred on the Coma Cluster. The 'body' of the stickman is not a physically linear structure, though; it is caused by the use of redshifts as ersatz distances. The large peculiar velocities in the cluster stretch out the redshift distribution in a line pointing towards the observer – the 'finger of God' effect. The stickman's arms represent parts of the 'Great Wall', which runs for about 200 Mpc through the Coma Supercluster region. Other surveys, such as that based on the IRAS galaxy catalogue (which emphasises star-forming spiral and irregular galaxies at the expense of weakly FIR emitting ellipticals), confirmed this picture and we now have the 2dF and SDSS redshifts for hundreds of thousands of galaxies which delineate the 3-D distribution out to distances $\simeq$1 Gpc.

---

[4] They may have been more receptive to this idea, at the time, than US and other western astronomers because of the influence of the Soviet school of cosmologists headed by Zeldovich, who proposed that galaxy formation should occur in flattened 'pancakes' rather than in more spherically symmetric structures.

[5] In order to visualise the 3-D structure from 2-D plots, it is usual to divide the surveyed volume up into slices, or wedges, with a wide range in one coordinate on the sky, usually Right Ascension, but a narrow one in the perpendicular direction, that is Declination.

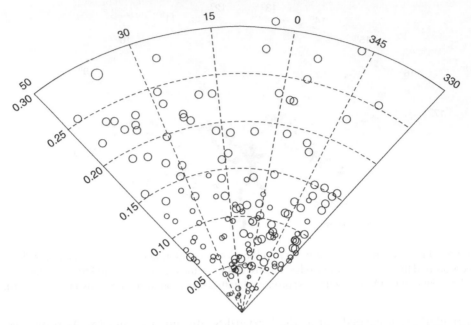

Figure 7.12 The large-scale distribution of X-ray selected galaxy clusters in the direction around the South Galactic Pole, out to a redshift z = 0.3 (from Cruddace *et al.*, 2002).

The 'bubbles' in the foam-like distribution are the voids, which can be 50 Mpc or more in extent. The most famous is probably that seen in the direction of the constellation Boötes, which may have a volume approaching a million cubic Mpc. Voids need not be entirely empty, but certainly show a very large under-density. The lack of mass inside them, compared to the average density outside, leads them to behave like 'negative' masses and the galaxies in or near the walls of the voids therefore move outwards, increasing the density contrast still further.

Superclusters, that is clustering of other smaller groups about a large cluster (as in the Virgo Supercluster), or more particularly the clumping of large clusters themselves, was probably first suggested by Shapley in 1933. Amongst other regions, he studied the large number of clusters seen in the general direction of the constellation Centaurus, which are nowadays known as the 'Shapley concentration'. This region is also known as the 'Great Attractor', as the whole set of local galaxies around us appears to be 'flowing' towards it, presumably because of the accumulated gravitational attraction of all the masses in that direction, as discussed in section 7.9.

7.8   Clustering statistics

From the irregularities noted in the 1930s, it was evident that the variance of the numbers of galaxies in different patches of the sky was larger than would be expected if galaxies were randomly distributed outside the rich clusters, and Zwicky quantified this through the wonderfully named 'index of clumpiness'. Hubble also discussed such 'counts in cells' and showed that for counts N, $\log(N)$ had a gaussian (or 'normal') distribution. Such a distribution of N is known in statistics as 'log normal'.

The search for underlying mathematical descriptions of the galaxy distribution was continued through the 1950s by both astronomers and statisticians, but it was not until the early 1970s that real progress was made. At this point Jim Peebles demonstrated the efficacy of the 2-point correlation function, a function closely related to those used in statistical mechanics. It had been used occasionally by previous observers, but was now developed into *the* key tool relating theory and observation.

7.8.1   Correlation functions

We can define the angular 2-point correlation function as follows. Suppose the mean number density of galaxies on the sky (down to some fixed magnitude limit) is $\overline{N}$. Now choose any random galaxy as a 'centre'. The probability δP of finding another galaxy in a small area $\delta\Omega$ placed a distance θ from this centre will obviously be $\overline{N}\delta\Omega$ if the galaxies are randomly distributed. However, if galaxies are clustered together on the scale θ (for instance if they come in groups of size greater than θ), then this probability will be increased; a given galaxy will be more likely to have a companion at the given distance. We therefore define the angular 2-point correlation function $w(\theta)$ through

$$\delta P = \overline{N}\left(1 + w(\theta)\right)\delta\Omega \qquad (7.8.1)$$

An alternative is to define w through the probability of finding galaxies in each of two area elements $\delta\Omega_1$ and $\delta\Omega_2$ a distance θ apart:

$$\delta^2 P = \overline{N}^2\left(1 + w(\theta)\right)\delta\Omega_1\delta\Omega_2 \qquad (7.8.2)$$

We can easily extend this definition to the 3-D distribution of galaxies in space, instead of just across the sky, if we consider the probability of finding a second galaxy in a volume element δV at a physical distance r from our chosen centre galaxy,

$$\delta P = \overline{n}\left(1 + \xi(r)\right)\delta V \qquad (7.8.3)$$

where now $\overline{n}$ is the mean volume density of galaxies and ξ is the spatial 2-point correlation function. We specify '2-point' as we can also consider the separations of 3 or more galaxies, using 'n-point' correlation functions.

Clearly w can be calculated directly from the positions of galaxies on the sky, for example, from photographic plates. If the observed number of pairs of galaxies which are between θ and $\theta + \Delta\theta$ apart is $N_p(\theta)$ and the corresponding number of pairs for a random distribution of the same number of galaxies over the same area of sky is $N_r(\theta)$, then a simple estimate of the correlation function is just

$$w(\theta) = \frac{N_p(\theta)}{N_r(\theta)} - 1 \qquad (7.8.4)$$

If we did not need to worry about any complications such as the edges of the surveyed region, then with N_g galaxies, N_r would evidently be $0.5\,N_g\overline{N}2\pi\theta\Delta\theta$ (the factor 0.5 is to avoid counting each physical pair twice). In practice, N_r is most easily calculated by directly setting down random distributions of points in what are called Monte Carlo simulations.

Until the 1990s, the number of galaxies with known redshifts, and hence distances, was very small compared to the number of galaxies whose positions were known (for instance there were 2 million galaxies in the APM galaxy catalogue but only 2000 in the contemporaneous redshift survey of IRAS galaxies). Thus, nearly all the earlier work used the angular function w. It was soon found, from many different samples with different depths (i.e. magnitude limits), that w could always be well fitted by a power law

$$w(\theta) = A\theta^{-\delta} \qquad (7.8.5)$$

with $\delta \simeq 0.8$ and the amplitude A depending on the survey. Specifically A decreases as the depth of the survey increases, because it depends on how much the clustering signal is 'washed out' by having structures at a range of distances projected on top of one another. It would be preferable, then, to have a measure of the physical clustering in 3-D, rather than its projection onto the sky.

Fortunately, as with projecting the volume density of, say, stars in a galaxy to provide its surface brightness, if we start with a power law spatial correlation function

$$\xi(r) \propto r^{-\gamma} \qquad (7.8.6)$$

then

$$w(\theta) \propto \theta^{1-\gamma} \qquad (7.43)$$

Thus the observed power law w can be interpreted as coming from a spatial function of the form

$$\xi(r) = \left(\frac{r}{r_0}\right)^{-1.8} \qquad (7.8.7)$$

where r_0 is called the correlation length. We can interpret r_0 as the scale at which density irregularities switch from being 'linear' (i.e. small compared to the overall

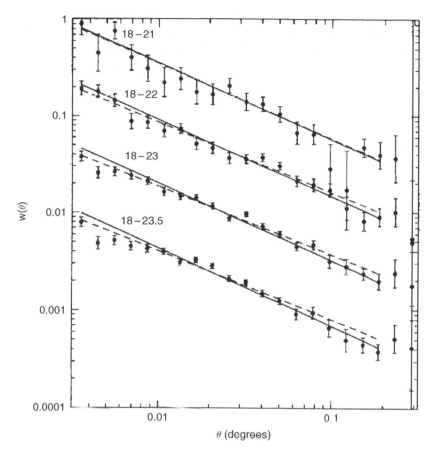

Figure 7.13 Estimates of the correlation function $w(\theta)$ for samples limited at different magnitudes (Reproduced with permission from Couch, Jurcevic and Boyle, 1993).

density) to 'non-linear', since evidently the 'excess' density $\bar{n}\xi$ is there equal to the density itself. The correlation function falls to close to zero when $r \sim 30$ Mpc.

7.8.2 Limber's formula

To see how r_0 can be related to the amplitude A for a particular survey, notice that having galaxies in two areas $\delta\Omega_1$ and $\delta\Omega_2$ θ apart on the sky is equivalent to integrating over all possible volume elements δV_1 and δV_2 (at distances x_1 and x_2, and separated by $r(x_1, x_2, \theta)$) in those two directions. Hence

$$\delta^2 P = \bar{N}^2 (1 + w(\theta))\delta\Omega_1\delta\Omega_2 = \int\int (1 + \xi(r))\bar{n}^2\Psi_1 \mathrm{d}V_1\Psi_2\mathrm{d}V_2$$

$$= \bar{n}^2 \int\int (1 + \xi(r))\Psi(x_1)x_1^2\mathrm{d}x_1\delta\Omega_1\,\Psi(x_2)x_2^2\mathrm{d}x_2\delta\Omega_2 \qquad (7.8.8)$$

The function Ψ here is the 'selection function', that is the fraction of all galaxies at distance x which are bright enough to be included in the survey under consideration (and hence is dependent on the LF). The integrals over the non-clustered parts are obviously equal (no clustering in 3-D implies none in 2-D), and

$$\overline{N} = \overline{n} \int \Psi(x)x^2 \mathrm{d}x \qquad (7.8.9)$$

We then have

$$w(\theta) = \frac{\int x_1^2\Psi(x_1)\mathrm{d}x_1 \int x_2^2\Psi(x_2)\xi(r)\mathrm{d}x_2}{\left(\int x^2\Psi(x)\mathrm{d}x\right)^2} \qquad (7.8.10)$$

This is 'Limber's formula'; a version which allows for the curvature of space in the distances and the volume elements is called the cosmological Limber's formula.

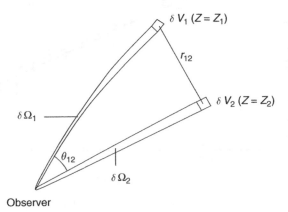

Figure 7.14 The geometry of the relationship between $w(\theta)$ and $\xi(r)$ (Reproduced with permission from Phillipps *et al.*, 1978).

If we set $x = (x_1 + x_2)/2$ and assume that the clustering scale is much less than the overall depth of the survey, then $x_1 \simeq x_2 \simeq x$, and if we are interested only in small angular separations

$$r^2 = (x_1 - x_2)^2 + x^2\theta^2 \qquad (7.8.11)$$

Taking ξ to be a power law as above, we can now rewrite the double integral for w as one integral over the distance x and one over the ratio of line of sight to tangential separations $u = (x_1 - x_2)/x\theta$, viz.

$$w(\theta) = \frac{\int \Psi^2(x)x^4\mathrm{d}x \int r_0^\gamma\left(x\theta(1+u^2)^{-1/2}\right)^{-\gamma}x\theta\mathrm{d}u}{\left(\int \Psi(x)x^2\mathrm{d}x\right)^2} \qquad (7.8.12)$$

This gives

$$w(\theta) = I_\gamma S(\Psi) r_0^\gamma \theta^{1-\gamma} \qquad (7.8.13)$$

where

$$I_\gamma = \int_{-\infty}^{\infty} (1 + u^2)^{-\gamma/2} du = \frac{\sqrt{\pi}\Gamma(\gamma/2 - 1/2)}{\Gamma(\gamma/2)} \qquad (7.8.14)$$

is a just a constant (Γ is the Gamma function we met in the integral of the Schechter function), and

$$S(\Psi) = \frac{\int \Psi^2 x^{5-\gamma} dx}{\left(\int \Psi x^2 dx\right)^2} \qquad (7.8.15)$$

depends only on the selection function.

For simple magnitude limited surveys, Ψ has a fixed shape and we can see that the amplitude of w, A, must vary with the characteristic depth D (say the distance at which M_* galaxies are at the survey limit) as $D^{6-\gamma}/(D^3)^2$, that is $D^{-\gamma}$. Alternatively, we can see that the galaxy surface density must vary as D^3, so

$$A \propto \overline{N}^{-\gamma/3} \qquad (7.8.16)$$

the so-called 'scaling relation'.

7.8.3   Clustering lengths

Studies at various magnitude limits demonstrate that for nearby galaxies, in general, $r_0 \simeq 7\,\text{Mpc}$ (for our chosen H_0). As we might expect, though, ellipticals are more clustered than spirals, thus having a larger value of r_0. This also means that FIR-selected samples are less strongly clustered than optically selected ones. In the last few years, with the advent of the large redshift surveys, it has become possible to determine the spatial correlation function directly, confirming the earlier indirect results and demonstrating that more luminous galaxies are more strongly clustered than galaxies around L_*. Galaxy clusters are also (super)clustered according to a rather similar law, but with $r_0 \sim 30\,\text{Mpc}$. In this case, it is the richer clusters that are more clustered than the less populous ones. In addition, we could investigate the clustering of more distant objects, perhaps radio galaxies or quasars, but the interpretation of these results is bound up with the general evolution of structure in the universe, so we will leave this until Chapter 8.

As determined here, ξ obviously applies to the distribution of visible galaxies, not necessarily to the distribution of the underlying mass. Indeed, since they have different r_0, the galaxies and clusters cannot both trace the underlying mass, so there is the suspicion that neither might. The difference between ξ_g for the galaxies and ξ_m for the mass is known as the 'bias', i.e.

$$\xi_g = b^2 \xi_m \qquad (7.8.17)$$

One type of bias, first discussed by Nick Kaiser in 1984, is 'high peaks biasing', i.e. galaxies may only form where the density exceeds some particular threshold level. If fluctuations occur on a range of scales then it will be easier for a certain small-scale peak to rise above the threshold if it is on top of a larger (cluster) scale positive fluctuation. The high peaks, where the galaxies form, are statistically more clustered than the overall mass.

Correlation functions are not the only statistics that can be used to study the galaxy distribution. For instance, the 'void probability function' measures the probability of finding a void of a certain size. However, if we know all the n-point functions, then we have everything there is to know statistically about a point distribution; all other statistics are functions of these.

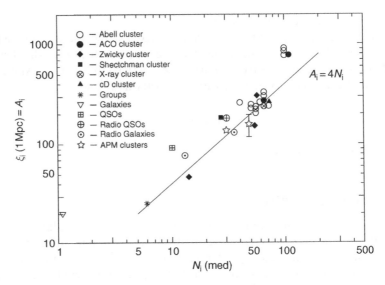

Figure 7.15 The variation of the correlation strength with cluster richness (Reproduced with permission from Bahcall and West, 1992).

7.8.4 The power spectrum

The most useful of the other measures is the power spectrum. This is related to ideas in signal processing theory and is the Fourier transform of the correlation function,

$$P(\boldsymbol{k}) = \int \xi(r) \exp(i\boldsymbol{k}.\boldsymbol{r}) \, d^3\boldsymbol{r} = 4\pi \int_0^\infty \xi(r) \frac{\sin kr}{kr} r^2 \mathrm{d}r \qquad (7.8.18)$$

If we think of the density distribution as a superposition of waves of different scales (as in the high peaks biasing idea), then $P(k)$ is measuring the contribution of waves of wavenumber k.

The power spectrum has the advantage of measuring large-scale (small k) fluctuations more accurately than does ξ, partly at least because the powers on different scales have uncorrelated errors. If the density at any point r is $\rho(r) = \rho_0(1 + \delta(r))$ then the variance of δ on scale R

$$<\delta_R^2> \simeq \frac{k^3 P(k)}{2\pi^2} \equiv \Delta_k^2 \qquad (7.8.19)$$

for $k \simeq 1/R$. The clustering strength in terms of the power on different scales is often quantified by the parameter σ_8 which is the average fluctuation on scales of $8h^{-1}$ Mpc (i.e. $\simeq 11$ Mpc for our assumed H_0). For most surveys σ_8 is found to be close to unity. We can also define the bias in terms of the power spectrum as

$$\Delta_g^2 = b^2 \Delta_m^2 \qquad (7.8.20)$$

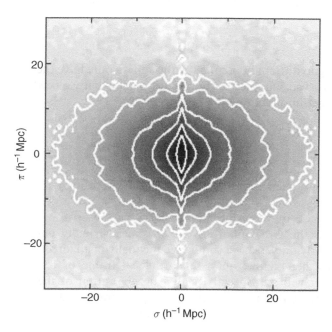

Figure 7.16 The distortion of the correlation function by the 'finger of God' effect and infall, as seen in the 2dFGRS data (Reproduced with permission from Hawkins *et al.*, 2003). The contours are lines of constant $\xi(\sigma, \pi)$.

7.9 Velocities

If we determine the 3-D correlation function in 'redshift space', that is by using the measured redshift as the third dimension (the usual case, of course, as few galaxies have independently determined distances), then as we saw with the 'stickman', peculiar velocities of galaxies will distort the clustering pattern. On small scales, the 'finger of God' effect will force radial separations to look larger than they really are, so if we decompose ξ into components tangential to and along the line of sight (often, if confusingly, denoted by $\xi(\sigma, \pi)$), contours of high ξ will be similarly stretched out in the radial direction.

However, on large scales, we can expect that in the vicinity of a dense region galaxies will be 'falling in'. This means that those on the far side will have smaller than expected redshifts and those in the foreground larger ones, thus reducing the apparent separations and squashing the low ξ contours. Since velocity effects are induced by the size of the mass fluctuations, they depend on the mean density ρ_0 and on $\xi_m = \xi_g/b^2$. Thus if we measure the distortion of the clustering ξ_g in redshift space, then we can estimate a combination of the mean density and the bias which turns out to have the form $\rho_0^{0.6}/b$.

Another way to use the peculiar velocities is to try to reconstruct the density field, regardless of the observed luminous matter. We know from standard dynamics that the velocity field $v(x)$, that is the velocities at each point x, must satisfy

$$v(x) = -\nabla \Phi(x) \tag{7.9.1}$$

where Φ is the gravitational potential. Although we only have the radial components ν_r from the observed Doppler shifts, this is exactly what we need to integrate along some line of sight in direction (θ, ϕ) to give Φ, i.e.

$$\Phi(s, \theta, \phi) = - \int_0^s \nu_r(r, \theta, \phi) \mathrm{d}r \tag{7.9.2}$$

Then we can, as usual, use Poisson's equation to obtain the density from the potential.

We can measure our own peculiar velocity via observations of the microwave background. As the CMB photons arose from a virtually uniform all pervading plasma in the early universe, we should see it equally in all directions (isotropy of the universe). This then defines our standard local frame of reference. However, a moving observer will see Special Relativistic aberration of the radiation and this results in us seeing different fluxes of energy from one direction on the sky to another. This is equivalent to a dipolar variation of the temperature across the sky

$$T(\theta) = T_0(1 + \nu \cos \theta/c) \tag{7.9.3}$$

where θ is the angle away from our velocity vector. Measuring $T(\theta)$ therefore allows us to determine our velocity relative to the standard frame.

Allowing for the orbit of the Earth and Sun around the Galactic Centre and for the motion of the Galaxy in the LG, the observed CMB dipole implies that the LG is moving at about $630 \, \mathrm{km\,s^{-1}}$ towards $\ell = 276°$, $b = 30°$. We might have expected to be falling towards the Virgo Cluster, but using the known distance (from Cepheids) and the mean cluster redshift, we find that the LG 'Virgocentric infall' is only $200–300 \, \mathrm{km\,s^{-1}}$. Our overall velocity is at a large angle to the Virgo direction.

By comparing redshifts to distances measured by other means, such as the T–F or F–J relations, we can determine peculiar velocities of other galaxies relative to the LG. These turn out to be quite small, 50–100 km s^{-1}, implying a 'quiet' local Hubble flow. Adding the two velocities we can find the velocities of other galaxies relative to their local frames. Remarkably it is found that a volume of size at least 60 Mpc is flowing coherently at ~600 km s^{-1} in the direction of the 'Great Attractor' or Shapley Concentration region.

If we add the gravitational forces from all the galaxies out to some distance, then as each contributes according to its M/D^2, their effects scale in proportion to their observed fluxes (assuming $M \propto L$). We can therefore estimate the gravitational 'dipole' or accumulated acceleration, simply by considering the distribution and brightnesses of galaxies on the sky (i.e. without knowing individual distances). Using IRAS galaxies (to minimise the effects of absorption and especially the Zone of Avoidance) it is found that there is reasonably good agreement on the directions of the gravitational dipole and the CMB dipole. The integration also implies that there is a significant contribution to the dipole from a surprisingly large range of distances; peculiar velocities are not just generated by a galaxy's close neighbours.

8 Galaxy evolution

8.1 Looking back

We have already introduced most of the main ideas relevant to galaxy evolution in previous chapters, for instance star formation histories, the ageing of stellar populations, chemical evolution and galaxy interactions. However, we have looked at these essentially by way of their end products, the current population of galaxies and their stellar and gaseous content. In this chapter we will instead consider how we can directly observe the process of galaxy evolution in progress.

Thanks to the finite speed of light, we see very distant galaxies as they were a light travel time ago. In an infinitely old universe which was unchanging on average (such as the old 'Steady State' universe of Bondi, Gold and Hoyle), we would see galaxies at all stages of their development in any given, sufficiently large volume of space, regardless of its distance from us. However, there appears to be no population of very young or forming galaxies locally; almost all seem to have some stars billions of years old. In any case, as we shall see below, the numbers of galaxies seen at different apparent brightnesses preclude such a universe. Given that the distance between galaxies is increasing (Hubble's law), the natural inference is that the universe as a whole is expanding. Running the expansion backwards, we deduce that the universe used to be 'smaller' and denser, and that a time $\sim 1/H_0$ ago all the matter in the universe should converge back to an infinitely dense state, the Big Bang.

For a finite age universe, galaxies will clearly be systematically younger at large look-back times, so we can identify galaxies at different distances from us as being at different stages of their evolution. Indeed, we might already guess that, by

The Structure and Evolution of Galaxies   Steven Phillipps
© 2005 John Wiley & Sons, Ltd

extrapolation of Hubble's (relatively local) redshift-distance law, we should be able to judge look-back time from observed redshift. However, for the huge distances involved, if we want to look far back in time, we need to account properly for the structural properties of the universe as a whole. Thus in order to determine how time and redshift should be related, we need to venture into the area of cosmology.

8.2   Redshift and distance

Let us first consider in more detail how the redshift of a galaxy is associated to the expansion of the universe in general. Space-time in relativity is governed by its 'metric', which measures the 'distance' (in the general mathematical sense) between two neighbouring points. In Special Relativity, for 'events' separated by coordinate differences dt, dx, dy, dz in 4-D space-time, the metric is

$$ds^2 = c^2 dt^2 - dx^2 - dy^2 - dz^2 \tag{8.2.1}$$

which we could also write in terms of spherical polar coordinates as

$$ds^2 = c^2 dt^2 - dr^2 - r^2 d\theta^2 - r^2 \sin^2 \theta \, d\phi^2 \tag{8.2.2}$$

In GR the metric, that is, the local geometry of space, is determined by the distribution of mass, but on the largest scales it must be consistent with the Cosmological Principle. The most general metric which satisfies spatial homogeneity and isotropy is the Robertson–Walker metric

$$ds^2 = c^2 dt^2 - a^2(t) \left(\frac{dr^2}{1 - kr^2} + r^2 d\theta^2 + r^2 \sin^2 \theta \, d\phi^2 \right) \tag{8.2.3}$$

Here r, θ and ϕ are called co-moving coordinates, as a galaxy's position in these coordinates does not change (apart from any local motion) as the universe expands, and t is the cosmic time. Finally k is a number, either $+1$, 0 or -1, which describes the 'curvature' of the universe.

The radial coordinate r does not represent an actual distance, though. As we can see from the metric, we have to combine it with $a(t)$, called the 'cosmic scale factor', which accounts for the expansion (i.e. galaxies get further apart with time). Specifically, the locally measured distance between two neighbouring points at the same θ and ϕ at time t must be $a(t) \, dr/(1 - kr^2)^{1/2}$. If we add together all these increments (which is like using a whole set of rulers placed end to end) along the line from a distant source at coordinate r_1 to an observer at the origin, we evidently have at time t,

$$D_p = a(t) \int_0^{r_1} \frac{dr}{(1 - kr^2)^{1/2}} \tag{8.2.4}$$

This is called the 'proper distance' to the source. Using standard integrals, we can see that it is given by $a(t)\sin^{-1}r_1$ for $k=1$, $a(t)r_1$ for $k=0$ or $a(t)\sinh^{-1}r_1$ for $k=-1$. This is a fundamental distance, in the sense that it is explicitly defined from the metric, but having to arrange for the long string of local observers is not terribly practical!

Before trying to construct a more observational definition of distance we need to consider how our idea of redshift relates to the expansion factor a. By definition, photons travel along paths in space-time for which $ds^2=0$, called 'null geodesics'. For a purely radial path, this requires

$$c\,dt = -a(t)\,\frac{dr}{(1-kr^2)^{1/2}} \tag{8.2.5}$$

(negative because r decreases as t increases). We can therefore integrate along the path of a photon emitted from a source at r_1 at time t_1 and arriving at an observer at the origin at t_0, to obtain

$$\int_{t_1}^{t_0}\frac{c\,dt}{a(t)} = \int_0^{r_1}\frac{dr}{(1-kr^2)^{1/2}} \tag{8.2.6}$$

But if a second photon (or wavecrest) is emitted at time $t_1+\delta t_1$ and received at $t_0+\delta t_0$ then

$$\int_{t_1+\delta t_1}^{t_0+\delta t_0}\frac{c\,dt}{a(t)} - \int_0^{r_1}\frac{dr}{(1-kr^2)^{1/2}} = \int_{t_1}^{t_0}\frac{c\,dt}{a(t)}$$

which implies

$$\frac{c\,\delta t_0}{a(t_0)} + \int_{t_1+\delta t_1}^{t_0}\frac{c\,dt}{a(t)} = \frac{c\,\delta t_1}{a(t_1)} + \int_{t_1+\delta t_1}^{t_0}\frac{c\,dt}{a(t)}$$

i.e.

$$\frac{c\,\delta t_0}{a(t_0)} = \frac{c\,\delta t_1}{a(t_1)} \tag{8.2.7}$$

We can identify $c\delta t$ with the wavelength of the photon so

$$\frac{\lambda_0}{a(t_0)} = \frac{\lambda_1}{a(t_1)} \tag{8.2.8}$$

where λ_1 and λ_0 are the emitted and received wavelengths. So, by the definition of redshift,

$$1+z = \frac{\lambda_0}{\lambda_1} = \frac{a(t_0)}{a(t_1)} \tag{8.2.9}$$

Thus the photon's wavelength is stretched by the same factor as that by which the universe has expanded during its time of flight (the look-back time to the source). This gives the required relationship between the observable redshift and the expansion.

Notice that since a is a function of t, this provides a parameter

$$H = \frac{\dot{a}(t)}{a(t)} \qquad (8.2.10)$$

to describe the rate of expansion. If we consider the proper distance between two galaxies,

$$D_p = a(t)f(r_1) \qquad (8.2.11)$$

(where f denotes the possible functional forms noted above), then the rate at which they are moving away from each other, the proper velocity, will be

$$v_p = \dot{D}_p = \dot{a}f(r_1) = D_p\,\frac{\dot{a}}{a} = H D_p \qquad (8.2.12)$$

Thus H is the ratio of proper velocity to proper distance; so, as these reduce to the normal meanings of velocity and distance on small scales, H must be just what we have called Hubble's constant. However, we can now see that it may well vary with time, so it is better to refer to it as the Hubble parameter, with H_0 being its present day value.

8.2.1   Observable distances

To relate the redshift to observable distances and look-back times, though, we still require both some practical definitions of distance and a way of determining a as a function of t. Suppose that we work backwards from a simple measurement of distance. For instance, from simple trigonometry, we would expect an object of proper size δy at a distance D to subtend an angle $\delta\theta = \delta y/D$. But the angular part of the metric implies $\delta y = a(t_1)r_1\delta\theta$, where t_1 was the time at which the light we see left the source. We can define the 'angular diameter distance' D_A as

$$D_A = a(t_1)\,r_1 = (1 + z)^{-1}a(t_0)\,r_1 \qquad (8.2.13)$$

so that the standard trigonometric identity does apply. Notice that in the (simple) case $k = 0$, D_A is $(1 + z)$ times smaller than the proper distance.

Now consider, instead, how bright objects should look if placed at different radial coordinates. If we have a source of absolute luminosity L at r_1, at time t_1, emitting photons of energy $h\nu_1$ at a rate n_1 per unit time, then photons of energy $h\nu_0 = h\nu_1/(1 + z)$ will pass through a sphere of area $4\pi a^2(t_0)r_1^2$ at a rate $n_0 = n_1/(1 + z)$ at some later time t_0. Thus according to an observer at $r = 0$, $t = t_0$, the flux from the source will be

$$F = \frac{h\nu_1}{(1 + z)}\frac{n_1}{(1 + z)}\frac{1}{4\pi a^2(t_0)r_1^2} = \frac{L}{4\pi(1 + z)^2 a^2(t_0)r_1^2} \qquad (8.2.14)$$

But if we assumed the normal inverse square law to apply, then we should expect $F = L/4\pi D^2$, so we can define a 'luminosity distance' D_L via

$$D_L = (1 + z)\, a(t_0)\, r_1 \tag{8.2.15}$$

in order that this does hold. This distance is now *larger* than the angular diameter distance by a factor $(1 + z)^2$.

For small z, that is, for $a(t) \simeq a(t_0)$ (which we will shorten to a_0 for convenience) or $t \simeq t_0$, we can make a Taylor series expansion of $a(t)$ about the time t_0. Thus

$$a(t) \simeq a_0 + \dot{a}(t - t_0) + \frac{1}{2}\ddot{a}(t - t_0)^2$$

$$= a_0\left[1 + H_0(t - t_0) - \frac{1}{2}q_0 H_0^2(t - t_0)^2\right] \tag{8.2.16}$$

where we have defined H as above and, a new quantity, the 'deceleration parameter' by

$$q_0 = -\frac{\ddot{a}_0 a_0}{\dot{a}_0^2} = -\frac{\ddot{a}_0}{H_0^2 a_0} \tag{8.2.17}$$

Since $z = a_0/a(t) - 1$ we can invert the expression for $a(t)$ to give

$$z \simeq H_0(t_0 - t) + (1 + q_0/2)H_0^2(t_0 - t)^2 \tag{8.2.18}$$

or

$$\Delta t \equiv t_0 - t \simeq H_0^{-1}\left[z - (1 + q_0/2)z^2\right] \tag{8.2.19}$$

This is our first approximation to the relationship between look-back time and redshift which we anticipated above.

We can also obtain an approximate relation between redshift and the distances which we have defined. To start with, we know that the function $f(r_1)$ is just r_1 to second order, so we can set

$$r_1 \simeq \int \frac{c\, dt}{a(t)} \tag{8.2.20}$$

or

$$a_0\, r_1 \simeq c \int (1 + z)dt \simeq cH_0^{-1}\int (1 + z)[1 - (2 + q_0)z]dz$$

$$= cH_0^{-1}\left[z - (1 + q_0)z^2/2\right] \tag{8.2.21}$$

so, for instance,

$$D_L \simeq cH_0^{-1} \left[z + (1 - q_0)z^2/2 \right] \qquad (8.2.22)$$

This, like all the other distances we have defined, then reduces to just

$$D = cz/H_0 \qquad (8.2.23)$$

to first order, matching the usual form of Hubble's law.

Notice particularly that we have got this far merely by assuming the Cosmological Principle. However, in order to obtain more exact relationships and ones applicable at larger z, or to relate the cosmological parameters H_0 and q_0 to the actual content or structure of the universe, we need to add dynamical information about $a(t)$. This can come only by a consideration of Einstein's equations of GR, as they apply to the universe as a whole.

8.3   Cosmological models

Without going into the details of GR, we can nevertheless deduce the relevant equations for the development of $a(t)$, just from a simple Newtonian discussion of a uniformly expanding system of particles, due originally to Bill McCrae. Consider the universe as a continuous fluid with density ρ and pressure P, and look at the motion of a particle a distance x from the observer at the origin. Now from Newton's own work, we know that for a spherically symmetric system (we are assuming isotropy, of course), the matter in shells further from the origin than x do not contribute to the motion of the particle so its gravitational acceleration will depend only on the interior mass $M = 4\pi\rho x^3/3$ and we will have

$$\ddot{x} = -\frac{GM}{x^2} = -\frac{4\pi}{3}G\rho x \qquad (8.3.1)$$

But for a uniformly expanding system, all distances must scale according to

$$x(t) = x_0 \, a(t)/a_0 \qquad (8.3.2)$$

So we can treat equation (8.3.2) as one for the scale factor a, i.e.

$$\frac{\ddot{a}}{a} = -\frac{4\pi}{3}G\rho \qquad (8.3.3)$$

If we think now about the energy (per unit mass) of the particle, then the kinetic part is just $\dot{x}^2/2$, while the potential is $-GM/x$. By the same argument as above, we then have

$$\dot{a}^2 - \frac{8\pi}{3}G\rho a^2 = \text{constant} \qquad (8.3.4)$$

Writing the constant as $-kc^2$, which clearly has the correct dimensions,

$$\frac{(\dot{a})^2}{a^2} + \frac{kc^2}{a^2} = \frac{8\pi}{3} G\rho \tag{8.3.5}$$

Notice that if $k=0$ the kinetic and potential energies balance, as in the ordinary expression for escape velocity. Thus in this case the expansion will just coast to a stop at infinity. If the total energy is positive (k negative), then the expansion is unbounded; while if the energy is negative (k positive), the expansion will eventually turn round, and the particle will fall back in.

In fact, if we go through the whole machinery of GR, using Einstein's field equation to relate the amount of mass and energy present in the universe to its curvature, then we find that for the general case we should include a term for the pressure, as well as the density, in our equation for $\ddot{a}$ and that an additional constant term, the 'cosmological constant' Λ, is also allowed. This can be thought of as representing an extra energy density ('dark energy') not associated with any sort of matter. Thus we finally obtain the Friedmann equations (independently derived by Friedmann and Lemaître in the 1920s)

$$\frac{\ddot{a}}{a} = -\frac{4\pi}{3} G\left(\rho + \frac{3P}{c^2}\right) + \frac{\Lambda}{3} \tag{8.3.6}$$

and

$$\frac{(\dot{a})^2}{a^2} + \frac{kc^2}{a^2} = \frac{8\pi}{3} G\rho + \frac{\Lambda}{3} \tag{8.3.7}$$

The detailed treatment also demonstrates, as we have implied by our choice of symbol, that the k in the Friedmann equations is the same as that in the Robertson–Walker metric.

8.3.1   The density parameter

We can also derive a 'conservation equation'

$$\frac{\mathrm{d}\left(\rho a^3 c^2\right)}{\mathrm{d}t} = -P\frac{\mathrm{d}\left(a^3\right)}{\mathrm{d}t} \tag{8.3.8}$$

(in our Newtonian representation this would be Boyle's law relating the change in energy to the 'PdV work' done by the expansion), but this is actually a combination of our two previous Friedmann equations (8.3.6 and 8.3.7), so only two of the three are independent. If the pressure is negligible, as it is for ordinary matter at the present epoch, the conservation equation implies, as we would expect, $\rho \propto a^{-3}$, or more specifically (again using the subscript zero to represent present day values)

$$\rho(t)\, a^3(t) = \rho_0\, a_0^3 \tag{8.3.9}$$

On the other hand, if we look at the radiation content of the universe, then besides the decrease in the number density, each individual photon has an energy decreasing as $1/a$, so overall the energy density in radiation, ρ_r, falls as $1/a^4$, that is, faster than the mass density. Since we know from ordinary thermodynamics that radiation density[1] depends on temperature as $a_R T^4$, we must also have $T \propto 1/a$. Working backwards, it is clear that whatever the present ratio of rest mass energy of matter compared to energy in radiation (i.e. in the CMB), if we go to small enough a (large enough z), the radiation would have dominated. Given the present measured values, this would have occurred at redshifts above a few thousand, when the universe was at several thousand K or more. We will neglect the present day contribution by radiation in what follows.

Now define the present 'density parameter'

$$\Omega_0 = \frac{8\pi G \rho_0}{3 H_0^2} \tag{8.3.10}$$

and the contribution by the cosmological constant by a corresponding

$$\Omega_\Lambda = \frac{\Lambda}{3 H_0^2} \tag{8.3.11}$$

If we use this in the Friedmann equation (8.3.7) then

$$\dot{a}_0^2 = a_0^2 H_0^2 = \Omega_0 H_0^2 a_0^2 + \Omega_\Lambda H_0^2 a_0^2 - kc^2 \tag{8.3.12}$$

i.e.

$$kc^2 = H_0^2 a_0^2 (\Omega_0 + \Omega_\Lambda - 1) \tag{8.3.13}$$

Thus $k = 0$ corresponds to $\Omega_0 + \Omega_\Lambda = 1$, and larger (or smaller) values of $\Omega_0 + \Omega_\Lambda$ give positive (or negative) curvature. In the case of the $k = 0$ model with zero cosmological constant, the density is equal to the 'critical density'

$$\rho_c = 3 H_0^2 / 8\pi G \simeq 1 \times 10^{-26} \, \text{kg m}^{-3} \tag{8.3.14}$$

For other models

$$\Omega_0 = \rho_0 / \rho_c \tag{8.3.15}$$

In these terms, the observed mass in stars in galaxies, which we can obtain by multiplying the observed luminosity density $\mathcal{L}$ by a reasonable stellar population M/L ratio,

---

[1] We use a_R for the radiation constant to distinguish it from the scale factor.

corresponds to a stellar contribution to the baryon density $\Omega_b(\text{stars}) \simeq 0.005$. In fact, we can determine quite accurately what Ω_b should be, from the theory of Big Bang nucleosynthesis. Fusion of protons to form deuterons and helium nuclei requires a certain particle energy and hence temperature (that attained in the centres of stars). We can extrapolate our cosmological models backwards to predict that these conditions should have held at $t \sim 10^2\,\text{s}$ after the Big Bang. The amount of D or He produced is then dependent on the particle (hence mass) density at this epoch, so if we can observe or deduce the primordial D or He abundances then we can get at ρ_b. We also know ρ_r at the same epoch from the required temperature, so we can obtain the baryon-to-photon ratio. As this is conserved during the expansion, it will still have the same value today ($\simeq 6 \times 10^{-10}$), so this and the measured energy density in the CMB give us the present baryon density.

To get the primordial element abundances needed for this procedure, we can look at either primordial objects directly (i.e. at high z) or ones that we think have changed rather little over time. In the first case, we can observe the D abundance relative to II in QSO absorption line systems. Locally, we can look at galaxies with very low metal contents, as they will have processed their ISM relatively little. Since the He content (Y) should rise in step with the metal content (Z) during processing in stars, if we observe galaxies with low Z and extrapolate back their Y values to $Z = 0$, then we should obtain the primordial He abundance as required. The most recent results (as well as observations of CMB fluctuations, as discussed in section 8.3.2, which also depend on Ω_b) now give consistent values close to $\Omega_b \simeq 0.04$. Notice that this is much larger than the stellar value and implies a large contribution from baryons in interstellar and intergalactic gas at various temperatures.

8.3.2   The cosmic scale factor

Returning to our cosmological models, we can relate the deceleration parameter to the density since (from equation 8.3.6 for a matter dominated universe, with $P = 0$)

$$\frac{\ddot{a}_0}{a_0} = -\frac{4\pi}{3}G\rho_0 + \frac{\Lambda}{3} = -\frac{1}{2}\Omega_0 H_0^2 + \Omega_\Lambda H_0^2 \tag{8.3.16}$$

implying

$$q_0 = \frac{\Omega_0}{2} - \Omega_\Lambda \tag{8.3.17}$$

In particular, $q_0 = 1/2$ for the $k = 0$, $\Lambda = 0$ model. This is called the Einstein–de Sitter universe and, until recent observational developments, was long thought to provide the most likely and most reasonable model. We will still use it below as an example, as it gives the mathematically simplest results.

Continuing with the $P = 0$ case, since $\rho \propto a^{-3}$, we can also use our definitions of Ω_0 and Ω_Λ and equation (8.3.13) for kc^2 in the Friedmann equation (8.3.7) to obtain

$$\dot{a}^2 = H_0^2 \Omega_0 \frac{a_0^3}{a} - H_0^2 a_0^2 (\Omega_0 + \Omega_\Lambda - 1) + H_0^2 \Omega_\Lambda a^2 \tag{8.3.18}$$

Rearranging, this gives

$$\dot{a} = H_0 a_0 \left(\Omega_0 \frac{a_0}{a} + (1 - \Omega_0 - \Omega_\Lambda) + \Omega_\Lambda \frac{a^2}{a_0^2} \right)^{1/2} \tag{8.3.19}$$

which can be integrated to provide our required relation between a and t to describe the expansion. And since $a_0/a = (1 + z)$, it also gives us the relationship between the observed redshift of a source and the cosmic time at which it is observed.

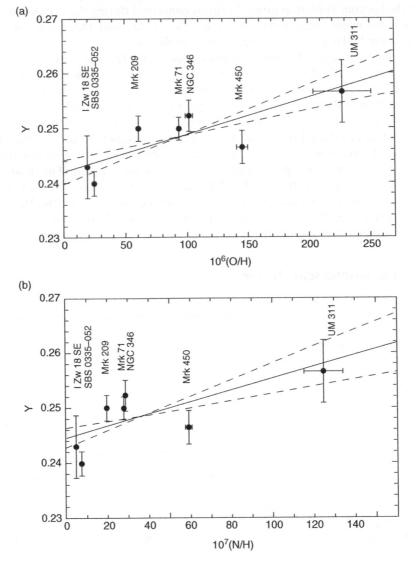

Figure 8.1   The variation of helium abundance (Y) with metallicity for metal poor galaxies (Reproduced with permission from Izotov and Thuan, 2004).

Unfortunately, the integral is not analytic in general cases, so must be solved numerically. It is analytic – but still not usually simple – if $\Lambda = 0$, and in the very simplest case of the Einstein–de Sitter universe, it reduces to

$$\frac{\mathrm{d}a}{\mathrm{d}t} = H_0\, a_0 (a_0/a)^{1/2} \tag{8.3.20}$$

giving

$$a = a_0 (3H_0 t/2)^{2/3} \tag{8.3.21}$$

or

$$3H_0 t = 2(1 + z)^{-3/2} \tag{8.3.22}$$

This also provides the standard result for such a model that the present age of the universe is

$$t_0 = \frac{2}{3}\frac{1}{H_0} \tag{8.3.23}$$

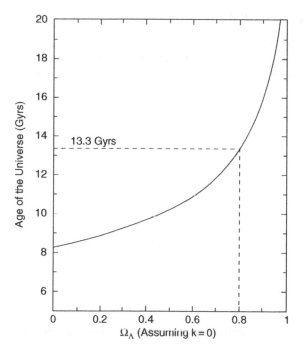

Figure 8.2 The age of the universe as a function of cosmological parameters for flat universes (Reproduced with permission from Driver, Windhorst, Phillipps and Bristow, 1996). The figure uses $H_0 = 80\,\mathrm{km\,s^{-1}\,Mpc^{-1}}$, so $1/H_0 \simeq 12.2\,\mathrm{Gyr}$.

The result, $a \propto t^{2/3}$, is also the asymptotic form for all other matter dominated models at early times (since then the term $\Omega_0 a_0/a$ is the dominant one in the general expression). Thus the results for this model provide a useful working 'rule of thumb' for consideration of cosmological effects. In particular, we can relate observational redshifts to the cosmic time at which the light was emitted in such a model (equation 8.3.22).

In all zero Λ models there is the simple limit on the age of the universe $H_0 t_0 \leq 1$, with the age decreasing as Ω_0 increases, obviously reaching $H_0 t_0 = 2/3$ for $\Omega_0 = 1$. However, as we can see from Figure 8.2, obtained by integrating the general formula (8.3.19), this is not the case for non-zero Λ models, which can have arbitrarily long ages as $\Omega_\Lambda \to 1$.

Observationally, the temperature fluctuations in the CMB across the sky can be characterised by a fluctuation spectrum $P(k)$, just as for the galaxy distribution discussed in section 7.8. Features seen in this spectrum, called acoustic peaks, occur at wave numbers, or angular scales, which depend on the cosmological parameters. Current evidence from the Wilkinson Microwave Anisotropy Probe (WMAP) experiment, which produced its first results in 2003, especially when combined with observations of high redshift supernovae, described below, strongly indicates a flat universe but with a large contribution from the cosmological constant. The current 'concordance model' has $\Omega_0 = 0.3$, $\Omega_\Lambda = 0.7$. In this model, $t_0 = 13.7\,\mathrm{Gyr}$ for our standard value of H_0.

8.3.3   Distance revisited

It still remains to relate the distances we have defined to redshift. We can start by finding the radial coordinate r as a function of redshift, for instance for the simple Einstein–de Sitter case. The equation of a null geodesic is then

$$dr = \frac{c\,dt}{a(t)} = \frac{c\,da}{a\dot{a}} = \frac{c\,da}{H_0\,a^{1/2}a_0^{3/2}} \qquad (8.3.24)$$

i.e.

$$a_0\,r_1 = \frac{2c}{H_0}\left[\left(\frac{a}{a_0}\right)^{1/2}\right]_{t_1}^{t_0} \qquad (8.3.25)$$

where t_1 and t_0 are the times of emission (from a galaxy at r_1) and reception of a photon, as before. Thus

$$a_0\,r_1 = 2c\,H_0^{-1}\left[1 - (1+z)^{-1/2}\right] \qquad (8.3.26)$$

We can then use this to write down expressions for the observable distances, for instance

$$D_L = (1+z)a_0\,r_1 = 2c\,H_0^{-1}\left[(1+z) - (1+z)^{1/2}\right] \qquad (8.3.27)$$

The more general form of (8.3.26) for any zero Λ model is called the Mattig equation; no analytic form exists for general Λ.

8.4   The Hubble diagram

We have already seen in equation (8.2.22) that we can write the luminosity distance to second order in z as $D_L = cH_0^{-1}(z + (1 - q_0)z^2/2)$, so if we can use our standard trick of finding a class of 'standard candles' of fixed absolute luminosity, then by measuring apparent brightnesses (magnitudes) for such objects at a range of redshifts we should be able to use the shape of the plot of m against z to determine q_0. This plot, called the Hubble diagram, is one of the classical cosmological tests for the curvature of the universe; distances are larger, and hence magnitudes fainter, for smaller values of q_0. We can, of course, also do this with the precise (or at least numerically computed) form of D_L for general Ω_0 and Ω_Λ and hence attempt to determine these cosmological parameters.

The original Hubble diagram test was used on whole galaxies for many years, but with no convincing results. However, in the past decade it has become possible to detect supernovae at ever increasing distances. Type Ia SNe[2] are especially good standard candles. In fact, this can be improved still further by measuring the time-scale of the fall-off in brightness after the peak of the output, since this correlates with the precise value of the maximum brightness. Using Type Ia SNe at redshifts up to, and now beyond, $z = 1$ groups such as the Supernova Cosmology Project and the

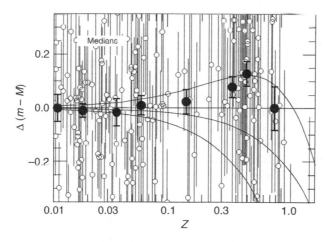

Figure 8.3   The Hubble diagram for Type Ia SNe (Reproduced with permission from Tonry *et al.*, 2004). The $(m - M)$ values are plotted relative to the prediction for an $\Omega_0 = \Omega_\Lambda = 0$ model and the upper curve is for the concordance model. The solid symbols are medians over the indicated z ranges.

---

[2] These are due to mass transfer from a binary companion onto a white dwarf carrying it over the Chandrasekhar limit for stability and thus causing it to collapse.

High z Supernova Search team have demonstrated that q_0 must be negative; high z supernovae are too faint for any standard Friedmann–Lemaitre universe. The best fit curve therefore requires a non-zero Λ and combining the SN evidence with the CMB constraints (which imply that the universe is very close to being flat, i.e. $\Omega_0 + \Omega_\Lambda = 1$) we arrive at the concordance values of Ω_0 and Ω_Λ, which give $q_0 \simeq -1/2$.

The first serious attempt to measure q_0 in this way had been by Humason, Mayall and Sandage in the 1950s. They used the first-ranked (brightest) galaxies in rich clusters, out as far as $z = 0.2$, as their putative standard candles. The magnitudes plotted must of course be corrected for effects such as Galactic absorption, galaxy k-corrections and how much of the galaxy's light is actually included in the measurement (for instance because of the low surface brightness of the outer parts, especially at high z, or because of the use of a fixed angular-sized aperture). In addition, there may be effects on the accuracy of the standard candle assumptions, including the Scott effect (if the LF does not have an upper cut-off, we must inevitably find brighter galaxies as we survey the larger volumes out to higher z), the richness effect (if galaxies are selected from a statistical LF, richer clusters should have brighter first-ranked members) and the Bautz–Morgan (B–M) effect (the magnitude of the first-ranked galaxy would be expected to vary with the B–M type of its cluster, which effectively measures the difference between the first and second brightest objects).

8.4.1   Evolutionary corrections

However, even above these effects, there is the question of whether first-ranked galaxies at different z should even be expected to have the same M. In fact, we might expect that galaxies will change their luminosities systematically with time, and hence z, through evolutionary processes. Without going into too many details, we might foresee three main possibilities, some of which we have introduced in previous chapters. First, any set of stars formed at some specific epoch will rapidly fade in total brightness as the bright blue, short-lived, massive stars come to the end of their lifetime. They will then fade further, but more slowly, as more and more stars evolve off the main sequence, even though they can become red giants (recall section 3.4). Second, though probably less important for the first-ranked cluster galaxies, which will be massive ellipticals, there may be ongoing, but variable amounts of, new star formation. And third, a galaxy may accumulate extra stars, or gas to make them from, via merging or accretion.

If we allow a general evolutionary rate dL/dt, where t is our usual cosmic time, then we can see from our previous approximations for D_L and $t(z)$ that we will have a relationship between apparent and absolute luminosity (of a galaxy as it would be at the present day) of the form

$$
F = \frac{L_0 H_0^2}{4\pi c^2 z^2} [1 + (q_0 - 1)z] \left[1 - \frac{1}{L_0} \frac{dL}{dt}(t_0 - t) \right]
$$

$$
= \frac{L_0 H_0^2}{4\pi c^2 z^2} \left[1 + \left(q_0 - \frac{1}{H_0 t_0} \left[\frac{d\ln L}{d\ln t} \right]_{t_0} - 1 \right) z \right] \qquad (8.4.1)
$$

(where we have used equation (8.2.18) to go from $(t_0 - t)$ to z), implying an obvious change in the deduced q_0 in the presence of luminosity evolution.

For an old stellar population, from equation (3.4.5), d ln L/d ln $t = -1 + \theta x \simeq -0.6$. Also $H_0 t_0 \simeq 1$, so the implied change in the apparent q_0 from the evolution of the stellar population will be $\Delta q_0 \simeq -1/2$. However, giant elliptical cluster galaxies are also likely to gain mass with time, giving a positive d ln L/d ln t. If, for example, accretion onto the central galaxy varies as t^κ, with κ theoretically expected to be of order 0.5, then d ln L/d ln $t \simeq f_0 \kappa$, where f_0 is the fraction of the present mass which has been added in this way. For growth by major mergers, this could be almost unity, so $\Delta q_0 \simeq +1/2$. As both of these changes were large, uncertain and in opposite directions, it is not really surprising that no definitive value of q_0 came out of this test. Most of the other 'classical' tests, presented in the equally classic 1961 paper by Allan Sandage on cosmological tests possible with the new 200″ telescope on Mount Palomar, are equally subject to these difficulties.

However, we can reverse this and use the same sorts of measurements to empirically constrain the amount of evolution of the galaxies. For instance, assuming the concordance cosmology, first-ranked cluster galaxies appear to show a net slightly

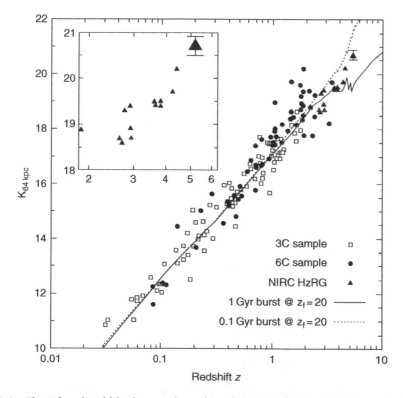

Figure 8.4   The *K*-band Hubble diagram for radio galaxies (Reproduced with permission from van Breugel *et al.*, 1999). The inset shows a set of data from the 10 m Keck Telescope while the lines in the main panel are predictions for two evolutionary models of galaxies formed at very high *z*.

'negative' evolution, in the sense that they were somewhat less luminous at earlier times. Since they contain old stellar populations which must have been brighter in the past, we can infer that the stellar mass was smaller than in present day first-ranked galaxies, at a level that turns out to be consistent with the dynamical predictions in such a universe. A similar story appears to apply to ellipticals in general. Generally speaking, the observational evidence, for instance looking at the fundamental plane or colour–magnitude relation at different epochs, seems to favour a combination of stellar fading and mass increase (both by a factor $\sim$2) which results in little overall change since $z \sim 1$, at least.

We should note that we can also look at the Hubble diagram for other specific galaxy types. Following the identification, in 1960, of the radio source 3C295 with the brightest galaxy in a cluster at $z = 0.46$, powerful radio galaxies became prime targets for high redshift galaxy studies. Optical identification of virtually all the 3CR (Revised 3rd Cambridge) catalogue radio sources in the early 1980s gradually increased the furthest redshifts sampled from around 1 to 1.8. Study of sources at lower flux limits, from later Cambridge surveys and surveys carried out elsewhere, has led to the discovery of radio galaxies out to $z \simeq 4$ or 5.

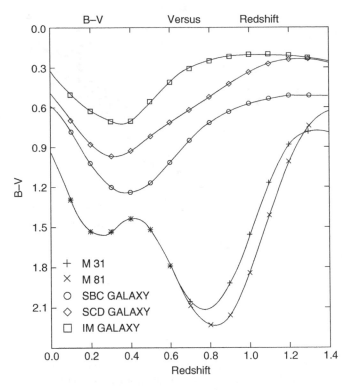

Figure 8.5 Galaxy colours as a function of redshift (Reproduced with permission from Coleman, Wu and Weedman, 1980).

Especially if we look in the infra-red *K* band, their Hubble diagram is remarkably tight (and virtually linear, in log *z*), all the way from local galaxies at *z* = 0.01 out to the most distant radio galaxies observed. Again this presumably reflects the play-off of stellar evolutionary effects against changes in stellar mass, and also suggests a rather high redshift of formation, in order to generate such similar properties for all radio galaxies at a given epoch.

In fact, we can track the stellar populations, fairly independently of the mass accumulation, by going back to the galaxy colours. Giant ellipticals (including those hosting radio galaxies) seem to follow the predictions of 'passive evolution' due to simple fading of an old stellar population. The existence of 'red' elliptical galaxies even at *z* = 2 suggests a short-lived burst of star formation at redshift at least 3 to 5. The correlation between black hole mass and galaxy or bulge velocity dispersion is taken to imply concurrent formation, the growth of the black hole probably being fuelled by infalling gas during spheroid formation. Thus the existence of quasars, and presumably therefore black holes, at *z* up to 6 probably implies the existence of massive spheroids at these very early times, less than 1 Gyr after the Big Bang.

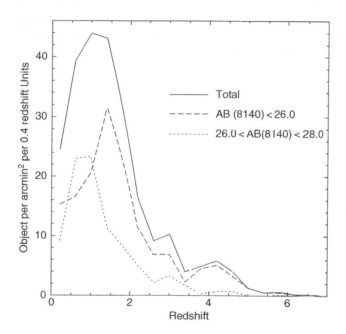

Figure 8.6 The photometric redshift distribution for HDF galaxies (Reproduced with permission from Fernandez–Soto, Yahil and Lanzetta, 1999).

8.5 Galaxy colours, photometric redshifts and LBGs

Stellar population evolution in general is also reflected in the changing observed colours of galaxies with redshift. These are due to a combination of the k-correction, as the spectrum is shifted across the observed bands, and the actual changes in the stars themselves. The latter can be predicted from the type and age of the stellar populations assumed to be present, as discussed in section 3.4.

Of course, this approach is not limited to the brightest ellipticals discussed in the previous section. From stellar models and a given SFH we can calculate the tracks through colour–magnitude or colour–colour diagrams for any galaxy type. Some standard curves are shown in Figure 8.5. Very sophisticated models now exist for tracking the evolution, including the chemical enrichment as star formation proceeds and the related variation in dust opacity, though they still have inherent uncertainties due to, for instance, remaining problems in stellar structure models and in the precise form (and universality) of the stellar IMF.

In recent years the use of colours has gained great prominence as the basis of the photometric redshift technique. By observing in several bands it is possible to find the best matching galaxy type and redshift to reproduce all the observed colours. Six or seven separate bands, preferably including the ultra-violet and infra-red, are found to give good discrimination and hence a good estimation of z, with errors in z as low as about 0.05, though one can make do with fewer colours at the expense of larger uncertainties. Although not giving precise distances, the photometric zs are sufficient to investigate, for instance, the overall distance distribution in a sample. The photometric redshift technique has been used especially widely on HST observations, particularly the HDFs, until recently the deepest ever optical observations, reaching $I \simeq 28$.[3]

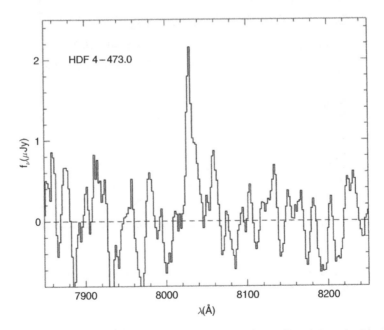

Figure 8.7   The spectrum of the $z = 5.6$ galaxy HDF 4-473.0 (Reproduced with permission from Weymann *et al.*, 1998) showing Lyα emission at 8029 Å.

---

[3] The Hubble Ultra Deep Field (UDF) data, released in 2004, reach about 1.5 magnitudes fainter than the HDFs.

Photometric *z*s, confirmed by spectroscopic follow-up, indicate the presence of galaxies out to $z = 5.6$, with most between $z = 0.5$ and 2. However, the types of galaxies appear to change significantly above $z \sim 1.5$. Both the optical morphologies (as revealed by the superb resolution of HST) and the best-fitting spectral templates indicate that at faint magnitudes and high *z*s there is an increasing fraction of irregulars, with a shortage of real spirals.

8.5.1　Drop-outs

The discovery of extremely high redshift galaxies, from amongst the vast numbers of faint images visible in HST data, is facilitated by use of the so-called 'drop-out' technique developed by Chuck Steidel and colleagues. This was first used to discover the 'Lyman break galaxies' (LBGs) at redshifts around 3. The spectrum of any source drops away sharply bluewards of the Lyman break, the end of the Lyman series of spectral lines at 91.2 nm, because any such short-wavelength photons can be absorbed in photoionising any hydrogen they encounter. Thus when, at *z* around 3, this break is redshifted past the *U* band (i.e. to around 360 nm), we expect the *U–B* colour to become very red or the object to be completely undetected at *U*. At progressively higher *z* the break will move across the other bands at longer wavelengths and the object will drop out of samples selected at *B*, *V* or *R*. In addition, at higher *z*, the accumulated absorption from the Lyman alpha forest lines (section 8.7) blankets out the flux shortward of Lyα in the source galaxy. Thus *R* band drop-outs, for example, occur at *z* above 4.8, and *I* band drop-outs have recently been used to search for galaxies at redshifts above 5.6. The nature of the LBGs themselves is still not completely established; though the numbers and clustering are consistent with them being the direct ancestors of present-day giant ellipticals, another view is that they could be small galaxies which will later merge with others to form a current epoch type galaxy. The galaxies seen in deep HST data at redshifts 4 and 5 (though presumably the most luminous in existence at that time) appear to be relatively small compared to present day galaxies, with half-light radii of 0.2″ to 0.4″, corresponding (in the current best guess cosmology) to less than 3 kpc.

8.5.2　FIR and sub-millimetre sources

Even if they are not 'drop-outs', as such, we may be able to select out other very high redshift objects by choosing those with extreme colours. In particular, 'Extremely Red Objects' (EROs), are usually defined to have optical to infra-red colours $R–K > 5$. In fact, EROs consist of at least two separate classes, elliptical-like galaxies at redshifts above 2 (often having signs of merging when studied in detail) and very dusty disc galaxies at redshifts around 1 to 1.5.

We should also mention here, the sub-millimetre detection of distant galaxies, using instruments such as SCUBA. Far from making them disappear at a given wavelength, increasing redshift can actually boost the brightness of galaxies in the sub-mm region. Since the typical dust-emission spectrum for a starburst galaxy peaks at around 100 μm, redshift moves the peak towards the observed band, giving rise to a negative k-correction and counteracting the dimming by distance.

Thus, in principle, a $z = 10$ source could be almost equally bright as one at $z = 1$. Observed high z sub-mm sources are plausible candidates for elliptical galaxies in formation, or at least during a major star-forming episode. Their dust emission fluxes suggest shrouded starbursts with an SFR $\sim 1000 M_\odot$ per yr. This would be sufficient to produce all the stars in a giant elliptical in a few $\times 10^8$ yr. Their space density also appears consistent with that expected for the precursors of modern day ellipticals.

8.6   Number counts

The observed redshift distribution $n(z)$, either spectroscopic or photometric, tells us how many galaxies are bright enough to be detected at each z. This must be due to both the actual number present and the fraction bright enough to be seen, and so will depend on the luminosity evolution of individual galaxies and on any number evolution if galaxies are created or lost.

Clearly related to this are galaxy number counts (also known as number–magnitude counts), where one merely considers the number of galaxies visible down to various magnitude limits. In the simplest case of a uniform distribution (number density n_0) of nearby galaxies, we can easily demonstrate the expected form. Suppose first that all galaxies have the same absolute luminosity L_i. The number of galaxies apparently brighter than flux F will be the same as the number nearer than distance

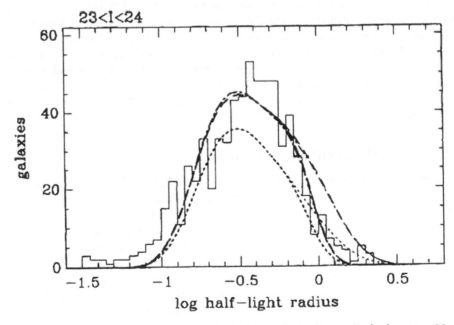

Figure 8.8   The distribution of half-light radii for faint galaxies (I magnitudes between 23 and 24) observed with HST (from Roche *et al.*, 1997). The curves superimposed on the data histogram show the predictions for various models.

$D_i = (L_i/4\pi F)^{1/2}$. The number seen in a survey covering a solid angle A on the sky will then be

$$N(>F) = \frac{A}{3} D_i^3 = \frac{A}{3} \left(\frac{L_i}{4\pi F}\right)^{3/2} \propto F^{-3/2} \qquad (8.6.1)$$

However the counts are usually expressed to magnitude limits, so converting from fluxes via $m = \text{constant} - 2.5 \log F$, we obtain

$$N(<m) \propto 10^{0.6m} \qquad (8.6.2)$$

But the luminosity L_i entered only into the constant term, so summing over arbitrary classes of galaxies (i.e. over all luminosities) we still get $N(>F) \propto F^{-3/2}$ and $N(<m) \propto 10^{0.6m}$.

In fact, in a cosmological context, number counts are much more informative than this, though more difficult to interpret. Taking the difficulties first, there is, for example, the purely technical point that, as we do not measure the bolometric luminosity, we have to allow for the k-correction to the observed flux in any band for higher redshift objects. This means that F no longer depends just on $1/D_L^2$. Further, D_L itself is a more complex measure of distance, due to curvature effects, so the volume out to D_L is not proportional to D_L^3. Indeed, from the R–W metric we can see that the proper volume element (the one measured by a local observer) which subtends a solid angle $A = \sin\theta\, d\theta\, d\phi$ is

$$dV = a^3(t) \frac{r^2 dr}{(1 - kr^2)^{1/2}} A = a^2(t) r^2 A c\, dt \qquad (8.6.3)$$

which a little manipulation using equation (8.3.7) shows to be equivalent to

$$dV = A \frac{D_L^2}{(1 + z)^5} \frac{c\, dz}{H(z)} \qquad (8.6.4)$$

where

$$H(z) = H_0 \left[\Omega_0(1 + z)^3 + (1 - \Omega_0 - \Omega_\Lambda)(1 + z)^2 + \Omega_\Lambda\right]^{1/2} \qquad (8.6.5)$$

At small z this reduces to the simple $A D_L^2 dD_L$ that we would expect, since $D_L \simeq cz/H_0$.

8.6.1 Evolution again

The above form makes it clear that measuring the number of galaxies to some magnitude limit – again if they were all equally bright – would be the same as counting all those in a volume which depends on the cosmological parameters. In the real case where we have a distribution of luminosities, given by the LF $\phi(L)$ or more generally a redshift-dependent proper number density $\phi(L,z)$, the number of sources in an interval dL and a volume element $dV(z)$ will be

$$d^2N = \phi(L,z)\,dL\,dV(z) \tag{8.6.6}$$

The total number observed to be brighter than F is therefore

$$N(>F) = \int_0^{L_{max}} \int_0^{z_L} \phi(L,z)\,dV(z)\,dL \tag{8.6.7}$$

where z_L is the maximum redshift at which a galaxy of luminosity L is brighter than F, that is $D_L(z_L) = (L/4\pi F)^{1/2}$.

In the case where galaxy numbers are conserved, we can set

$$\phi(L,z) = \phi_0(L[1-e(z)])(1+z)^3 \tag{8.6.8}$$

where ϕ_0 is the present-day LF and $e(z)$ is the fractional decrease in luminosity of galaxies (equivalent to a shift in magnitude) since epoch z. (We should also include the k-correction in e if we are observing in a fixed band, since this changes the luminosity we actually see.) To get an idea of the importance of the different terms, we can write $N(>F)$ to second order in z (using our previous approximations) as

$$N(>F) = \int_0^{L_{max}} \int_0^{z_L} A\left(\frac{c}{H_0}\right)^3 z^2[1-(2+2q_0+\nabla e)z]\,dz\,\phi_0(L)dL \tag{8.6.9}$$

where we have described the effect of the evolution through the gradient

$$\nabla e = \left[\frac{d\ln\phi}{d\ln L}\frac{de}{dz}\right]_{z=0} \tag{8.6.10}$$

(Note that ∇e will be negative since ϕ increases towards lower L.) But reversing the previous working for D_L we can also write z approximately as

$$z = \left(\frac{LH_0^2}{4\pi c^2 F}\right)^{1/2} - \frac{1}{2}(1-q_0)\left(\frac{LH_0^2}{4\pi c^2 F}\right)^2 \tag{8.6.11}$$

If we insert this in $N(>F)$ and integrate over z, we then obtain the slightly complicated, but informative, relation

$$N(>F) = \frac{A}{3(4\pi F)^{3/2}} \int_0^{L_{max}} \phi_0(L) L^{3/2} \left[1 - (3 + 4\nabla e) \left(\frac{LH_0^2}{4\pi c^2 F} \right)^{1/2} \right] dL \qquad (8.6.12)$$

This shows several things. First it reduces to the simple $F^{-3/2}$ law when F is large (i.e. for bright apparent magnitude local counts), but will exhibit curvature due to the second-order term. Second, the counts are strongly weighted to luminous objects (via the $L^{3/2}$ term), unless the LF is very steep. Third, to this order the q_0 dependence has cancelled out, but there *is* a dependence on the evolution parameter. Thus, as was first made clear by Beatrice Tinsley in the early 1970s, until we reach faint enough magnitudes that we are sampling to quite high redshifts, evolutionary effects dominate over the cosmology. It is for this reason that Hubble – who was (once again) the first to use number counts for cosmological purposes – found that *no* standard cosmological model fitted his data (since galaxy evolution was then unknown).

Hubble himself counted some 44 000 galaxies to the impressive limit of $B \simeq 21$ (where there are about 400 galaxies per square degree, at $z \leq 0.4$) using photographic plates from Mount Wilson. This was only surpassed in the late 1970s with deeper photographic studies on much larger telescopes and then CCD observations. The most recent studies have used HST imaging, especially in the HDFs where there are

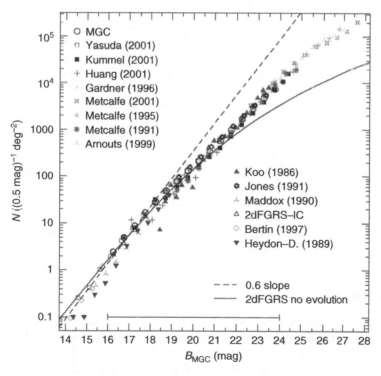

Figure 8.9 Galaxy number counts in the B band compiled by Liske *et al.* (2003).

about 2 million galaxies per square degree down to $B \simeq 29$, $R \simeq 28.5$, $I \simeq 28$. At the faintest limits, this equates to 1 object per $2.5'' \times 2.5''$ box. The Next Generation Space Telescope, now named the John Webb Space Telescope (JWST), working in the IR, will be able to count so many galaxies that the images will be 'confusion limited', that is effectively overlapping.

At these faint magnitudes, our simple low z approximation breaks down, of course, and the counts are determined by a combination of the evolution and the cosmological parameters. If we believe that the latter are now well tied down by the WMAP and supernova observations, then we can return to Tinsley's techniques and try to use the counts to infer the evolution. This is not simple, though, as – even ignoring any technical difficulties in measuring faint images – different morphological types will evolve differently and number conservation can no longer be assumed.

8.6.2   Faint blue galaxies

Even by the time the counts reached $B \simeq 24$, it was apparent that there was an excess of galaxies, compared to the 'no-evolution' predictions. In particular the $\log N(m)$ slope was around 0.4, rather than the 0.3 expected. (This is much less than the canonical 0.6 due to k-corrections and curvature effects.) The obvious interpretation was that luminous galaxies, already detectable at large distances, were even brighter in the past, increasing the volume in which they could be counted. However, the first redshift surveys reaching to similar depths showed that we were not seeing galaxies at significantly larger z than in the no-evolution models. Numerous solutions to this 'faint blue galaxy problem' were suggested, including extra populations of low- to moderate-luminosity galaxies seen only at higher z ('disappearing dwarfs'). These would have either faded to near invisibility (perhaps as LSBGs) or been swept up in mergers by the present epoch.

With knowledge of the apparent morphologies obtained from HST imaging, it is apparent that the 'excess' problem was, in some ways, actually due to a *lack* of higher z luminous objects relative to a simple evolution model. As we have already noted, there appears to be a shortage of fully formed spirals at $z \geq 1.5$, but many more irregulars than expected. The faint blue excess is therefore mainly due to small but vigorously star-forming galaxies. The accretion/merger of neighbours then leads to higher mass objects at lower z, so counteracts the fading of the stellar populations (and decrease in star formation activity), mimicking an overall no-evolution model in terms of $n(z)$. We will return to the evolution of more recently star-forming galaxies in the following sections.

8.6.3   Radio source counts

Of course, we can count other objects besides optically visible galaxies. Indeed, in the 1960s it was the counts of radio sources as a function of limiting flux, S which provided decisive evidence against the Steady State theory. This theory held that the universe looks the same not only at all positions and in all directions, but also at all times (the 'Perfect Cosmological Principle'). Since it had a well-defined cosmology, with $q_0 = -1$, and no resort to systematic changes in the galaxy population with cosmic time (and hence observed redshift), it made a directly testable prediction for $n(S)$ which proved to be incompatible with observations.

As with the optical galaxy counts, counts of radio sources have been extended to ever fainter flux limits, but again as for the optical counts they are now used to study the evolution of the population of radio sources and the different types, and intrinsic powers, of the sources which dominate in different flux regimes.

8.6.4 Quasars

Quasars can also be counted, either in the radio or the optical, and of course we generally know the redshift of each individual source. Their co-moving number density[4] increases rapidly towards higher z. Today quasars are distinctly rare, of order 1 per cubic Gpc, compared to a few million L_* galaxies in the same volume. However, at $z \simeq 3$ this increases by a factor $\sim 10^2 - 10^3$, so there is one quasar for every 10^3 to 10^4 bright galaxies. Most bright present day galaxies could once have hosted a quasar (i.e. an active, accreting black hole) if the quasar phase lasted for only a few Myr. Numbers of quasars probably decrease again beyond $z \simeq 4$, but this range is still not very well sampled.

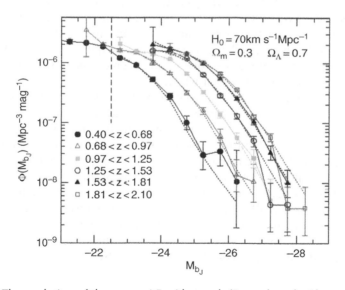

Figure 8.10 The evolution of the quasar LF with epoch (Reproduced with permission from Croom *et al.*, 2004).

---

[4] The co-moving density corresponds to the number of objects in a box which expands with the universe. With no creation or destruction of objects it should remain constant. The proper (i.e. locally measured) density, on the other hand, varies as $(1 + z)^3$ since the universe used to be smaller by that factor.

8.7 The night sky brightness

Closely related to the number counts is the question of the brightness of the night sky. As Hermann Bondi put it in his 1960 book *Cosmology*, the first ever cosmological observation was that it goes dark at night! If the universe were infinite, homogeneous, static (hence infinitely old) and full of stars and galaxies, then in whichever direction one looked, the line of sight would eventually intersect the surface of a star. The night sky brightness would therefore be the same as that of the surface of the Sun. That this is not the case is known as Olbers' paradox, though it had been discussed by other astronomers before Olbers in 1826. The solution, of course, is that one of the input assumptions must be wrong. Big Bang universes are not infinitely old, so if we look far enough then we will be seeing back to an epoch before stars had formed. These universes are not static either, of course, and in the case of the Steady State Universe (which *was* infinitely old), it was the expansion that solved Olbers' paradox, by redshifting the emitted photon energies to lower and lower values long before the point where the sky was filled with distant stars.

In fact, the true cosmological background brightness is much less than the observed night sky brightness. Even from a remote mountain-top observatory, the sky is dominated by more local contributions. These come from airglow in the atmosphere, zodiacal light (sunlight reflected off dust grains in the plane of the Solar System) and starlight in our Galaxy, including that scattered off interstellar clouds. A typical experiment to measure the true extragalactic background light (EBL) in the optical involves masking off or subtracting these other contributions. As the answer sought is the difference between two comparable numbers, it is still rather uncertain. In terms of the unit S_{10}, which is the surface brightness equivalent to the flux from one 10th magnitude A0 star spread over one square degree,[5] the typical night sky brightness at a dark site is about $90\,S_{10}$, of which $1 \pm 1\,S_{10}$ is EBL.

This is consistent with the direct summation of the light from visible galaxies. Taking the number counts in the blue down to $B = 26$ accounts for about $0.4\,S_{10}$. Though the count slope is still steep enough at this point that the value has not yet converged (a slope of 0.4 gives equal light from the galaxies in each magnitude bin), we can see that the counts must flatten or turn over at not much fainter magnitudes, otherwise the EBL constraint would be violated. This gives limits on the counts and more importantly on the evolution even beyond the point where we can see the individual galaxies.

This technique has actually been adopted more in the infra-red, where the FIR sky brightness puts significant constraints on how far back in time we can extrapolate the quite strong evolution that is usually deduced from the corresponding counts.

Theoretically, the sky brightness $\mathcal{B}$ must be given by

$$\mathcal{B} = \int_0^\infty F \mathrm{d}N(F) = \int_0^{z_{\max}} \int_0^{L_{\max}} \frac{L}{4\pi D_L^2} \phi(L, z)\, \mathrm{d}L\, \mathrm{d}V \qquad (8.7.1)$$

[5] A unit which surely could only have been invented by optical astronomers! In 'real' units it is equivalent to $\nu S_\nu = 7.6 \times 10^{-9}\,\mathrm{J\,m^{-2}\,s^{-1}\,str^{-1}}$.

For instance, for an Einstein–de Sitter universe with no evolution, so that $\phi \propto (1+z)^3$ and $1+z = (3H_0t/2)^{-2/3}$, we have

$$\mathcal{B} = \int_0^{L_{max}} L \mathrm{d}L \int_{t_f}^{t_0} \frac{\phi_0(L)}{1+z} c\mathrm{d}t = \frac{2c}{5H_0}\mathcal{L} \tag{8.7.2}$$

where $\mathcal{L}$ is the luminosity density introduced in Chapter 2. Dimensionally this equation is evidently sensible, as we are integrating the volume density along a column of length $\sim c/H_0$ to get a surface density. Numerically $\mathcal{L} \simeq 1.4 \times 10^8 L_\odot$ Mpc^{-3}, so the predicted sky brightness works out to be about $29.3\,\mathrm{B}\mu$ or about $0.25\,\mathrm{S}_{10}$. Evolution is therefore allowed (in this model) to enhance the calculated EBL (effectively the average $\mathcal{L}$) by a factor of a few.

The background light can also be studied at other wavelengths. We have already mentioned the FIR background above and the other area of particular interest has been the X-ray background (XRB). As there is very little foreground this can be measured fairly straightforwardly, but until recently, X-ray detectors lacked the spatial resolution to separate out individual faint sources. Gradual improvements increased the amount which could be ascribed to resolved images though, and Chandra has now effectively resolved the whole of the XRB flux into individual sources – diffuse emission from clusters, discrete emission from AGN (including optically obscured ones) and, at low levels, the flux from normal galaxies.

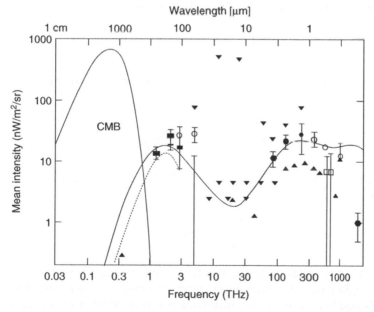

Figure 8.11 The extragalactic background light as a function of wavelength (Reproduced with permission from Wright, 2001).

8.7.1 Metal production

In the specific context of galaxy evolution, another interesting facet of the (optical) EBL is its relationship to the overall metal content of the universe. Both light and metals are produced primarily by the same stars in the same fusion processes. Thus, integrating the metal production rate over time gives essentially the same integral as in the EBL calculation. In fact,

$$\rho_{\text{metals}} \equiv Z\rho_b = \frac{4\pi}{\epsilon c}\mathcal{B} \tag{8.7.3}$$

where ϵ is the light emitted per unit mass of metals produced. For a standard Salpeter IMF, $\epsilon \simeq 0.5\,\text{J/kg/s/Hz}$ implying (at least for a flat spectrum object)

$$\rho_{\text{metals}} \simeq \mathcal{B} \times 10^{-30}\,\text{kg}\,\text{m}^{-3} \tag{8.7.4}$$

(with $\mathcal{B}$ in S_{10}). Compared to $\rho_b \simeq 0.04\,\rho_c \simeq 4 \times 10^{-28}\,\text{kg}\,\text{m}^{-3}$, $\mathcal{B} \sim 1$ implies a mean $Z \sim 10^{-2}$, consistent with direct estimates of the mean metallicity of the universe (and similar to the metallicity of the Sun).

8.7.2 QSO absorption lines

Finally, in relation to Olbers' paradox, we can consider the covering factor, that is how much of the sky is covered by galaxies (or other sorts of objects). If galaxies have a typical cross-sectional area σ and comoving number density $n(z)$, then the area of the sky covered by galaxies in a shell of volume $dV(z)$ will be

$$dA = n(z)(1+z)^3 \frac{\sigma}{D_A^2}dV \tag{8.7.5}$$

But the volume element can be written as $4\pi D_A^2 c\,dz/(1+z)H(z)$, so the covering factor

$$df_c = \frac{dA}{4\pi} = n(z)\sigma\frac{(1+z)^2\,c\,dz}{H(z)} \tag{8.7.6}$$

For the simple $\Omega_0 = 1$ case (where $H(z) = H_0(1+z)^{3/2}$) and no evolution this integrates to just

$$f_c = \frac{2}{3}\frac{c}{H_0}n_0\sigma_0\left[(1+z)^{3/2} - 1\right] \tag{8.7.7}$$

This is, of course, relevant to the absorption lines in quasar spectra produced when the quasar light passes through intervening gas clouds (section 4.6). Some of these, generally broad lines, are due to outflowing gas in the vicinity of the quasar, moving at relative velocities of several thousand $\text{km}\,\text{s}^{-1}$. Otherwise, the lines should trace the large-scale distribution of gas along the line of sight. Some will be in the metal line

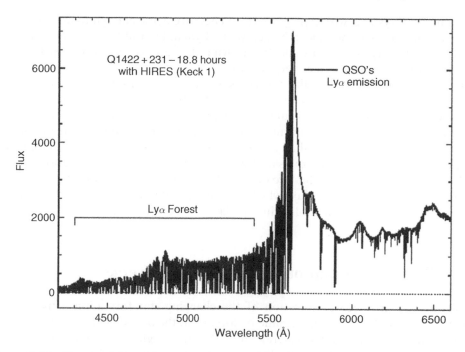

Figure 8.12   The spectrum of the quasar Q1422 | 231 taken with the Keck Telescope shows the huge number of absorption lines in the Ly α forest (Reproduced with permission from Pettini, 2001).

systems we mentioned earlier. As these are due to gas which has already been processed through stars, they are presumably associated with galaxies. The HI column densities range up to that seen looking through the central regions of a galaxy disc, $N_{HI} \geq 2 \times 10^{20}\,\text{cm}^{-2}$. These result in what are called 'damped' Lyman α lines, as the hydrogen layer is optically thick.

Given the cosmological factors we obtained above, it is straightforward to calculate how many absorption systems we should see along an average line of sight. If the comoving number density of absorbers is again $n(z)$ and their cross-sectional area is σ, then the differential count in a bin Δz wide will be

$$\Delta N = \frac{\mathrm{d}N}{\mathrm{d}z}\Delta z = \frac{\sigma n(z)c(1+z)^2}{H(z)}\Delta z \qquad (8.7.8)$$

Observationally $\mathrm{d}N/\mathrm{d}z \sim 0.2$ at $z \sim 3$ which requires very large absorbers with a radius around 50 or 100 kpc if n corresponds to the density of present day giant galaxies. Furthermore, we can see that for the simple $\Omega_0 = 1$ model with fixed n and σ we should have $\mathrm{d}N/\mathrm{d}z$ varying as $(1+z)^{1/2}$ whereas the observed numbers increase more like $(1+z)^2$, implying (even in other cosmological models) significant evolution in either number density or cross-section.

At lower column densities there are many more lines, the numbers rising as approximately $N_{HI}^{-1.7}$. Next down from the damped Ly α lines (DLAs) are the

'Lyman limit' systems with $N_{HI} \geq 2 \times 10^{17}$. These are dense enough to absorb essentially all photons with wavelengths below the Lyman limit at 91.2 nm (in their own restframe). Finally at levels around 10^{16} down to 10^{12} atoms per cm^2 we reach the Ly α forest, with vast numbers of very narrow lines.

Current models in fact see the absorbers as merely condensations in a ubiquitous IGM, rather than as discrete clouds; most of the IGM is ionised, so the absorbers represent pockets of denser neutral gas. The QSOALS then provide a useful route to calculating the number of ionising photons present at any epoch, an important cosmological quantity as we shall see in section 8.8. In the vicinity of a quasar, the extra ionising flux from the quasar itself will reduce the number of absorbers at redshifts close to its own – the 'proximity effect'. This reduction will depend on the ratio of the quasar's ionising flux (which we can calculate from its observed spectrum) to the ambient flux at that z, allowing the latter to be determined.

8.8   The star formation history of the universe

Probably the first secure indication of the changing star formation activity with epoch, at relatively recent times, came by looking at the overall galaxy population in rich clusters. We have noted several times that the bright galaxies in present day clusters are almost all early types, ellipticals and lenticulars, and these are, of course, red. At redshifts above 0.3, though, as was first noted by Harvey Butcher and Gus Oemler in 1978, there is a much higher fraction of blue galaxies. With the benefit of high resolution HST imaging, it appears that this Butcher–Oemler effect is due, at least in part, to a much higher fraction of spirals at the expense of S0s in the higher z clusters (though interpretation is complicated by the difference in galaxy populations between clusters of different density or richness and in the fractions of star-forming galaxies at different magnitude limits). Looking at it the other way round, it seems that over the past few Gyr most late type galaxies in rich clusters have been transmuted into early types through the cessation of star formation. Going back further, before the epoch corresponding to $z \simeq 1$ only luminous (massive) galaxies appear to be found in the 'red sequence'.

A general alternative to the colour or drop-out methods of searching for high redshift objects is to try to target actively star-forming objects. We know from section 4.12 that star formation is accompanied by the production of emission lines, and this provides us with two related techniques. Suppose we image some area of the sky (often one containing a known high redshift object such as a quasar) in two narrow bands (i.e. through filters which transmit only a narrow wavelength range). If one of the bands covers a prominent emission line (say Lyman α) at the relevant redshift, then there will be far more flux through this band than a neighbouring band off the line. Thus, if we tune our wavelength to Ly α at the redshift of the quasar, we will be able to find any strongly line-emitting neighbouring objects at the same z. Long slit spectroscopy, where the slit is placed so as to cover several neighbouring objects as well as a known high z one can work in the same way, as we may find spectral lines at the same wavelengths but spatially separated from the main emission. Typical star-forming galaxies at high z are around 3 magnitudes fainter (at K) than the radio galaxies we met in section 8.4, and are often virtually undetectable in their optical continuum (as opposed to line) emission. Searches around QSOs tend to find many more line emitters than do 'blank field' surveys, suggesting the presence of strong clustering even at high z.

Of course, we should be aware that samples selected in this way can be contaminated by sources emitting in a different line, say [OII], at a lower redshift. Conversely, not all galaxies may be strong line emitters even at early times when star formation would be expected to be most important. The distant HDF galaxy 4–473.0, for instance, appears to have an SFR of only about $10 M_\odot$ per yr, similar to a present day Sc. In addition, some local starbursting galaxies have little Ly α emission because of dust absorption. The latter suggests that we should also look for high redshift objects via their dust *emission*. As discussed in section 8.5, this will be redshifted into the sub-mm regime which has recently been opened up by new instruments.

8.8.1   Star formation models

If we can obtain detailed spectra, then the stage of evolution reached by an individual galaxy at any epoch can be explored by looking at, for instance, its position in a plot of $H\delta$ equivalent width against $(B–R)$ colour. Suppose we start with a large burst of star formation which is soon truncated. As the starburst proceeds, the galaxy will start with very blue colours and negligible $H\delta$ absorption, since its light will be dominated by the very highest mass stars. (It will also show a very low M/L ratio if this can be measured.) In the post-starburst phase, it will reach a maximum in $EW(H\delta)$ when it is dominated by A stars (so called $E + A$ galaxies) before converging to the usual fairly low $EW(H\delta)$ and red colours of a present day elliptical galaxy. On the other hand, if we have the fairly steady star formation of a spiral (disc), $H\delta$ never reaches particularly high levels, but it should correlate with the overall colour as both reflect the ratio of fairly recent to fairly old stars (see the hatched region in Figure 8.13).

In general, one might expect the star formation to decline with time as the available gas is used up, and a useful parameterisation of this is to assume an exponential decay in SFR with some characteristic decay time τ, that is SFR $\propto \exp(-t/\tau)$. τ will be short for galaxies undergoing a one-off burst of star formation – as usually assumed for ellipticals – with larger values of τ corresponding to longer-lived star formation, as seen in spirals. For a constant SFR, probably a reasonable approximation for late type spirals and irregulars, $\tau \to \infty$. These SFRs can be used in conjunction with an assumed IMF to calculate the colours of the galaxy at any epoch, as mentioned in section 8.5. For instance, the colour of a present day elliptical can be reproduced quite well if we assume that a burst with $\tau \sim 1$ Gyr occurred more than 10 Gyr ago. Spirals are consistent with τ in the range ~ 3 to 10 Gyr. Evidence from the distribution of disc star ages and metallicities in our Galaxy confirms that its past SFR was never much higher than its current value. However, the global variation of star formation with epoch is hard to determine from such observations, suggesting that more direct measurement of the star formation rate in galaxies at different z is required.

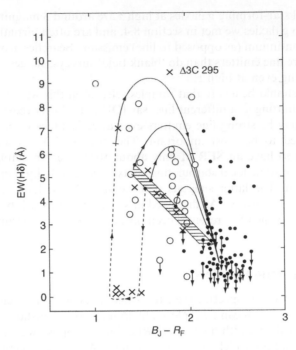

Figure 8.13 Evolution of starbursts in the $H\delta - (B–R)$ plane showing the position of 'E + A' galaxies (Reproduced with permission from Couch and Sharples, 1987).

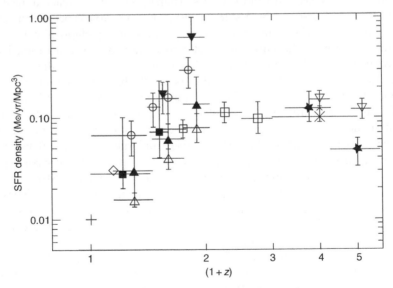

Figure 8.14 The SFH of the universe. Note that relative to the original Madau plot, the points have had a correction applied to try to allow for dust obscured star formation, thus removing the decline at high z (Reproduced with permission from a compilation of data by Mann *et al.*, 2002).

8.8.2   Star formation at high z

The 'Madau plot' utilises the number of strongly UV emitting galaxies seen in redshift surveys to determine the total SFR as a function of z. Stars producing the UV emission are mostly very short lived, and so give a direct measure of the SFR in the galaxy at the epoch at which it is observed. Thus the UV luminosity density should be proportional to the global star formation density, that is, the overall SFR per unit volume. Now at even moderate redshifts the rest frame UV flux is shifted into the visible, so we can readily determine the UV emission from visible galaxies. Observations from the Canada–France redshift survey in the mid 1990s were the first to show conclusively that there is a strong rise in SFR with look back time, an increase of a factor of order 10 by $z=1$ relative to that measured locally from Hα emission. Some of this decline by the present day will be due to the slowing down of SF in individual galaxies[6] and some, presumably, to the quenching of SF in denser environments as more and more galaxies fall into clusters.

The diagram was extended to higher redshift by using HDF data, and this suggested a decrease in the co-moving star formation density beyond redshift 2. However, recall that the SFR is being judged from the UV flux which is particularly susceptible to dust absorption. A (still rather uncertain) correction for this flattens out the SFR curve, implying fairly constant overall star formation from high z down to $z \sim 1$.

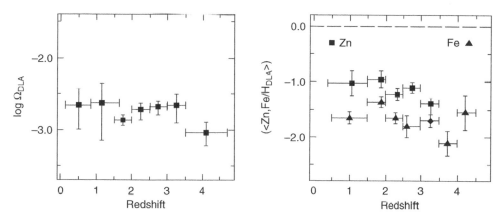

Figure 8.15   Evolution of the density in DLAs and of the DLA metallicity (Reproduced with permission from Pettini, 2001).

The deduced SFR(z) will clearly generate an increase in the total stellar density with time. One might therefore expect that the HI content of the universe at different epochs would show a corresponding decrease. If we take the DLAs (which account for most of the detected hydrogen) to have a number density $n_0 (1+z)^3$, radius r and column

---

[6] For instance, the 'downsizing' idea suggests that only lower mass objects have prolonged star formation.

density N_H, then the mass of a cloud will be $M \simeq \pi r^2 N_H m_p$. After a little algebra (and noting that $\rho_c \propto H^2$), the density parameter in gas at redshift z then becomes

$$\Omega_g(z) = \frac{n_0(1+z)^3 M}{\rho_c(z)} \simeq \frac{m_p N_{HI}}{\rho_c(0)} \frac{dn}{dz} \frac{H_0^2(1+z)}{c \, H(z)} \qquad (8.8.1)$$

where for simplicity we have ignored contributions from helium and molecular hydrogen. If we consider $z = 3$ and take $N_{HI} = 4 \times 10^{20} \, \text{cm}^{-2}$ and $dn/dz = 0.2$ we obtain $\Omega_g(z = 3) \simeq 10^{-3}$. The present day value (for gas in galaxies) is only $\simeq 10^{-4}$, but the mass in stars in spiral discs does correspond to $\Omega_b \simeq 10^{-3}$. This is therefore consistent with a picture where at high z almost all the baryons, now in stars, were in the interstellar (or intergalactic) medium. However, a more detailed look confuses this simple interpretation; recent determinations show rather little evolution in Ω_{DLA} across $z = 3.5$ to $z \simeq 1$. Assuming that the DLAs really are the progenitors of present day galaxies, this presumably implies that the gas used up is replaced (either in the same objects or newly formed ones) from the diffuse phase.

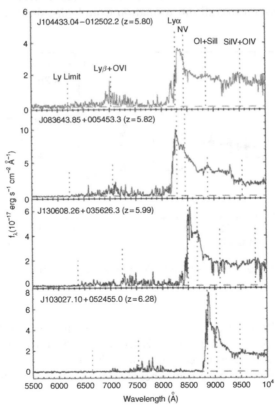

Figure 8.16   The spectra of high redshift quasars from the SDSS (Reproduced with permission from Becker *et al.*, 2001). The quasar at $z = 6.28$ shows an apparent total lack of flux just short of the Ly α emission, the first evidence for a Gunn-Peterson trough.

Likewise, measurements of the metal lines in QSOALS show a wide dispersion in metallicity but relatively little overall change with z. However, if we add in measurements of the metallicity of other types of object, QSOs themselves, the Ly α forest and Lyman break galaxies, then there does appear to be a strong dependence on environment, in that the densest systems have the highest Z at each z. Given that individual systems probably evolve through different 'categories' as they condense and then use up their gas to form stars, looking at an individual type of system, such as DLAs, may not be the best way to trace the overall build up of the global metallicity.

8.9   Reionisation and the first stars

We have seen (section 8.3) that the temperature of the radiation in the universe rises as we go back towards the Big Bang. Thus the (ordinary) matter in the early universe was completely ionised (and opaque to photons). When the universe cooled sufficiently, it became possible for atoms to form and the universe should have quickly become neutral – the epoch of 'recombination'. Nevertheless, observations of moderately high redshift objects such as quasars demonstrate that more recently the universe has been ionised again. If this were not the case, the neutral hydrogen in the intergalactic medium would absorb all the photons from the quasar shortward of the Lyman limit (in the cloud's rest frame) giving rise to a 'Gunn–Peterson trough' in the spectrum below 91.2 nm in the quasar's rest frame. The non-observation of this effect in quasars at redshifts up to ~5 puts severe limits on the fraction of neutral gas that can have been present at more recent epochs.

However, in observations of quasars at $z \simeq 6$ there are signs that the Gunn–Peterson effect has now been seen. This requires that enough luminous sources were present before this epoch – less than a Gyr after the Big Bang – to produce the necessary ionising radiation. The time when this occurred is called 'reionisation'. One possibility for the source of ionising photons is an early population of AGN, but the ones we observe tend to occur in large spheroids and appear to be relatively rare at $z \sim 4$ or 5. If they cannot provide the ionisation, then early generations of star-forming galaxies presumably do so instead.

The question of the earliest stars to form is a complex one. The very first stars (Population III) must have formed out of metal-free gas. This then prevents the gas cooling and collapsing by emission of photons in metal lines, in the way that normal proto-stellar clouds do, so cooling can only be via the lines of molecular hydrogen. Theoretical studies suggest that Population III stars will be preferentially of high mass, with many objects above $100 M_\odot$. These will have very short lifetimes but will produce copious ionising photons, as well as the metals which are present in the Population II stars which follow soon after. The high masses also explain why we see no surviving metal-free stars today.

8.10   Galaxy formation theory

So far we have said little about the actual formation of galaxies and how this might fit in to the general picture of structure formation in the universe. We know observationally that the CMB is remarkably smooth, the COBE satellite finding temperature fluctuations of only a few parts in 10^5. These fluctuations can be quantified through their power spectrum, which may have had the theoretically preferred Harrison–Zeldovich form, $P(k) \propto k^n$ with $n = 1$, at early times. They will grow with time and after the universe becomes matter dominated the corresponding small fluctuations in matter density will continue to grow via gravitational effects until they reach the level where they can 'collapse', that is separate out from the overall expansion of the universe. These density fluctuations will be primarily in the dark matter since this dominates over the baryons by a large factor. Given a primordial fluctuation spectrum, it is possible to calculate theoretically or numerically the growth of large-scale structure, which can then be compared to the distribution of high redshift objects such as LBGs, as well as to the present-day galaxy distributions, using the methods discussed in Chapter 7.

The same calculations also provide predictions of the numbers of rich (massive) clusters at any epoch, and on a smaller scale, the number of galaxies – or strictly dark matter halos – of different masses, in the so-called Press–Schechter formalism.[7] In the CDM models, including the consensus Λ dominated model, LCDM, the structure forms hierarchically or 'bottom-up', that is on small scales first. (Some previous models, such as the Zeldovich 'pancakes' mentioned in section 7.6, worked via fragmentation of large structures, so were of the 'top-down' variety.)

The collapse of the individual dark matter halos allows proto-galaxies to begin to form. However, at this point the simulation of purely gravitational growth becomes complicated by the hydrodynamical and other physical processes inherent in the formation of galaxies as such. Two basic theoretical approaches are possible. Either hydrodynamical processes can be modelled directly in the computer codes, or 'semi-analytic models' of the physical processes likely to be important during galaxy formation can be grafted on to run using the output from the pure N-body calculations (which generate the basic parameters and distribution of the dark halos).

The main ingredients of the semi-analytic models are a series of 'rules' specifying what happens to a given patch of the simulated system. They need to take account of the different behaviour of the dark matter particles, which will dominate the gravitational interactions but are otherwise non-interacting, and the baryonic particles, which are subject to dissipation, radiative energy loss, shocks and other such effects. In particular they require a prescription for how much cooling occurs and how the cooled gas turns into stars. Once stars have formed the models also need to account for the energy and mass which will eventually feed back into the interstellar medium via mass loss and supernovae. This feedback turns out to be one of the key (but least understood) determinants of, for example, the number of dwarf galaxies formed in the models. Another tricky point has been to make some, but not all of the gas cool, so as not to make excessively large stellar mass galaxies (the 'over-cooling problem'). In the most detailed models, the gradually changing metal content (which affects the cooling rate) and the production of dust (which

---

[7] The predicted shape of the halo mass function in Press–Schechter theory is the origin of the Schechter form of the LF.

affects the apparent brightness of the 'galaxies' formed) can also be taken into account. Despite the necessary simplifications, the semi-analytic models have had considerable success in matching both general large-scale structure in the universe and many properties of the observed galaxy population. Others features, such as the number of dwarf galaxies or the simultaneous matching of the shape of the bright end of the LF and the normalisation of the Tully–Fisher relation for spirals, have been more problematic, and point to the requirement for additional physics input to the models.[8]

The 'bottom-up' picture implies that large galaxies are built up from the hierarchical merger of smaller galaxies or sub-units. This is rather different from the older 'monolithic collapse' model wherein a large galaxy like our own would have formed from the collapse of a single large gas cloud. Also referred to as the ELS model, after Eggen, Lynden-Bell and Sandage, this envisages stars forming by fragmentation during the collapse (on a free fall time scale $\sim 10^8$ yr) to form the halo and globular clusters and eventually the bulge, but (for a spiral) allowing some of the gas to survive to fall to the central plane (not to a central point, because of rotation) where it shocks and dissipates energy. The gas can then remain in a disc configuration until further star formation occurs. There is actually some observational support for a picture like this, but this does not necessarily conflict with the hierarchical growth picture, as we could assume that mergers occur early (at least for some galaxies, such as proto-ellipticals), while the sub-clumps are still largely gaseous.

In general the merger of sub-units which have used up their gas supply will add to the central star mass, increasing the size of what will become the bulge. During this process, we presumably also have to grow the central supermassive black hole in step with the bulge mass. If there is unused gas around, this will gradually settle into a disc and (on a somewhat longer timescale) form further stars there, creating a disc component and hence a spiral type galaxy. However, another major merger might then disrupt the disc (as we saw in Chapter 3) turning the galaxy into an elliptical (which may or may not have the opportunity to grow another disc). Thus a galaxy can change its morphological type with time, perhaps explaining the apparently changing ratio of types with look-back time in the deep Hubble data. Different galaxies may also have widely different merger histories (described by what are called 'merger trees'), some perhaps arising from the merger of a few large clumps, others from the accretion of smaller clumps onto one existing large one.

One further quantity is important for disc systems, and that is their angular momentum. The angular momentum of a given halo will be built up as other proto-galaxies fly past, the 'tidal torquing' mechanism. Defining the importance of the angular momentum of a halo via the dimensionless 'spin parameter'

$$\lambda = J|E|^{1/2}G^{-1}M^{-5/2} \tag{8.10.1}$$

(where J is the total angular momentum, E the energy and M the mass of the halo), it is found that typical halos should have $\lambda \simeq 0.07$. High angular momentum should

---

[8] For instance, the inelegantly named 'squelching' of dwarf galaxy formation by ionising radiation.

result in a larger, therefore more diffuse disc, so the cosmic distribution of λ should be reflected in the observed distribution of disc surface brightnesses.

8.11   Further developments

Remarkable progress has been made in the last two decades in understanding the large-scale structure of the universe and the properties of galaxies both here now and in the distant past. We probably (though this is always a dangerous thing to claim) now have the general picture of how galaxies developed from initial fluctuations to their present state. Much of this progress has been brought about by technological advancement of observations and the theoretical work this has inspired. In the next decade, we can look forward to further progress. More data from WMAP and soon from the Planck satellite should measure the CMB fluctuations and thereby cosmological parameters to ever greater precision. The JWST will measure and map the high redshift universe to much greater depths even than the original Space Telescope and the Atacama Large Millimetre Array (ALMA) will probe the high z universe with high resolution at the wavelengths needed to uncover dust enshrouded star formation. Meanwhile, the GAIA satellite will measure the detailed properties of millions of stars in our own Galaxy, providing an unrivalled view of the (current) end point of galaxy formation and evolution.

Appendix: The magnitude system

Magnitudes

Following the convention of the Greek astronomers, most notably Hipparchus and Ptolomy, who divided the visible stars into those of the first magnitude, those of the second magnitude, and so on, optical astronomers measure the apparent (and, indeed, absolute or intrinsic) brightness of stars (and other objects) on a magnitude scale. Following the work of John Herschel, this was placed on a physical and quantitative footing in the mid-19th century by Pogson, who showed that the magnitude scale (i.e. the eye's response to light) was logarithmic and, moreover, that sixth magnitude stars were 100 times fainter than those of the first magnitude, that is, a difference of 5 magnitudes corresponded to a factor of 10^2 in flux. He therefore defined the (visual) magnitude such that

$$m = -2.5 \log_{10} F + \text{constant}$$

where F is the measured flux from the source and the constant is chosen so as to give $m \simeq 1$ for first magnitude stars. More precisely, on the standard 'Vega' system, the zero point is set so that the bright star Vega has a magnitude of 0.0.

Note that since magnitudes are logarithmic, a difference in magnitude $m_1 - m_2$ represents the ratio of fluxes. This means that, in this case, we do not need to worry about the constant above and have just

$$m_1 - m_2 = -2.5 \log_{10}(F_1/F_2)$$

or

$$\frac{F_1}{F_2} = 10^{-0.4(m_1 - m_2)}$$

The Structure and Evolution of Galaxies   Steven Phillipps
© 2005 John Wiley & Sons, Ltd

Bandpasses

Of course, the effect on the retina of the eye is due only to certain wavelengths of radiation, from blue to red or wavelengths ~400 to ~750 nm. The eye is most sensitive from 500 to 600 nm, the so-called visual band. If we observe, instead, with some astronomical detector, this will likely be sensitive to some other range of wavelengths (and with different efficiencies as a function of wavelength). This means that we must be careful to define the bandpass of the detector we are using to measure the flux. In general, the recorded brightness of a source will be

$$\int_{\lambda_{\min}}^{\lambda_{\max}} F(\lambda)S(\lambda)\mathrm{d}\lambda$$

where $F(\lambda)$ is the actual incident flux at wavelength λ, that is the energy flowing through unit area per second in a unit wavelength interval (measured, for instance, in $\mathrm{Wm}^{-2}\mathrm{nm}^{-1}$), $S(\lambda)$ is the efficiency with which the detector records photons of a particular wavelength, and $\lambda_{\min}$ and $\lambda_{\max}$ represent the range of detectable wavelengths (the detector bandwidth or bandpass).

Different bandpasses can be observed with different detectors or by observing through different filters. This allows us to define numerous magnitude systems such as the standard 'Johnson' UBV (ultra-violet, blue, visual) or the extended UBVRI (adding red and very near infra-red wavelengths). More recently the advent of the huge SDSS has led to the adoption of their ugriz (ultra-violet, green, red and two bands in the very near infra-red) as an alternative. Similar systems are used in the 'real' near infra-red (wavelengths above about 1 μ), the JHK bands being the most commonly used.

For a star like the Sun, most of the flux does emerge at wavelengths detectable by the eye, for instance, but there is actually a very wide spread of emitted wavelengths (stars are not far from the thermodynamic ideal of a 'black body' which emits or absorbs all wavelengths of radiation). Thus, a more encompassing definition is the 'bolometric' magnitude, which is the integral across all possible emitted wavelengths. The 'bolometric correction' (more often seen in stellar than extra-galactic astrophysics) is then the magnitude difference corresponding to the ratio of the visible flux to the total flux. For instance, the actual 'in band' blue flux from the Sun is $L_{\mathrm{B}\odot} = 5.2 \times 10^{25}$ W, compared to the bolometric 'standard' 3.9×10^{26} W.

Colours

Since the magnitude in a given passband represents how much, say, blue light is received, the difference between two magnitudes, B–R for instance, indicates the relative amounts of energy received at different wavelengths. This is known as a colour index, or colloquially just the 'colour' of a source. (It is easy to see that it indeed corresponds to what we mean by colour in everyday experience.)

Note that because of the 'negative' definition of magnitudes, a 'blue' object will have a small or negative B–R, while a 'red' object will have a larger positive B–R. We can, of course, define 'colours' which go beyond those visible to the eye, for instance infra-red J–K or an optical to infra-red index like B–H.

AB Magnitudes

In recent years, a somewhat more straightforward system of magnitudes has begun to be used. The 'AB' magnitude system essentially uses the same definition as the usual system, but without the normalisation to Vega, that is, at all wavelengths or frequencies it is a direct representation of $\log F$. Specifically, with F in $W\,m^{-2}\,Hz^{-1}$

$$m_{AB} = -2.5 \log F - 48.6$$

for any chosen band. The constant is such that m_{AB} does agree with the Vega system in the V band.

Absolute magnitude and distance modulus

As measured flux falls off as the inverse square of the distance D, the logarithmic nature of the magnitude system means that the apparent magnitude of a standard source will vary as $5 \log D$. In order to provide an absolute scale, one imagines

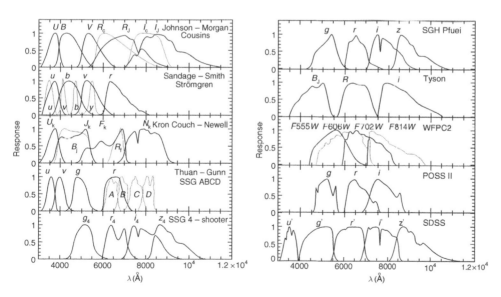

Figure A1   The bandpasses and sensitivities for a wide assortment of photometric systems (from Fukugita, Shimasaku and Ichikawa, 1995).

observing any star from a distance of 10 pc. The magnitude it would then have is called its absolute magnitude M and is clearly related to its observed apparent magnitude m by

$$M = m - 5 \log D/10$$

for D in parsecs. The difference

$$m - M = 5 \log D - 5$$

is called the distance modulus.

Figure credits

Figure 1.5 Hubble, 1925, The Astrophysical Journal, **62**, 409, The University of Chicago Press, Chicago.

Figure 1.6 Hubble and Humason, 1931, The Astrophysical Journal, **74**, 43, The University of Chicago Press, Chicago.

Figure 1.8 Perryman *et al.*, 1995, A&A, **304**, 69.

Figure 2.5 Sanchez-Portal *et al.*, 2004, Monthly Notices of the Royal Astronomical Society, **350**, 1087, Blackwell Publishing, Oxford.

Figure 2.6 Kregel, van der Kruit and de Grijs, 2002, Monthly Notices of the Royal Astronomical Society, **334**, 646, Blackwell Publishing, Oxford.

Figure 2.7 Schechter, 1976, The Astrophysical Journal, **203**, 297, The University of Chicago Press, Chicago.

Figure 2.9 Crampton and Georgelin, 1975, A&A, **40**, 317.

Figure 2.10 Schlegel, Davis and Finkbeiner, 1998, The Astrophysical Journal, **500**, 525, The University of Chicago Press, Chicago.

Figure 2.11 Fukugita, Shimasaku and Ichikawa, 1995, Publications of the Astronomical Society of the Pacific, **107**, 945, University of Chicago Press, Chicago.

Figure 2.12 Driver and Phillipps, 1996, The Astrophysical Journal, **469**, 529, The University of Chicago Press, Chicago.

Figure 2.13 Hardcastle *et al.*, 1997, Monthly Notices of the Royal Astronomical Society, **288**, 859, Blackwell Publishing, Oxford.

Figure 2.15 Karachentsev *et al.*, 2003, A&A, **398**, 479.

Figure 2.16 Dressler, 1980, The Astrophysical Journal, **236**, 351, The University of Chicago Press, Chicago.

Figure 3.1 de Vaucouleurs and Capaccioli, 1979, The Astrophysical Journal Supplement Series, **40**, 699, The American Astronomical Society, Washington DC.

Figure 3.2 Caon, Capaccioli and D'Onofrio, 1993, Monthly Notices of the Royal Astronomical Society, **265**, 1013, Blackwell Publishing, Oxford.

Figure 3.4 Graham and Guzman, 2003, The Astronomical Journal, **125**, 2936, The University of Chicago Press, Chicago.

Figure 3.5 Lauer *et al.*, 1995, The Astronomical Journal, **110**, 2622, The University of Chicago Press, Chicago.

The Structure and Evolution of Galaxies   Steven Phillipps
© 2005 John Wiley & Sons, Ltd

Figure 3.6 Nieto *et al.*, 1992, A&A, **257**, 97.

Figure 3.7 Naab, Burkert and Hernquist, 1999, The Astrophysical Journal, **523**, 133, The University of Chicago Press, Chicago.

Figure 3.8 Jacoby, Hunter and Christian, 1984, The Astrophysical Journal Supplement Series, **56**, 257, The American Astronomical Society, Washington DC.

Figure 3.9 Kennicutt, 1992, The Astrophysical Journal Supplement Series, **79**, 255, The American Astronomical Society, Washington DC.

Figure 3.10 Iben, 1967, Annual Review of Astronomy and Astrophysics, **5**, 571, Annual Reviews, Paolo Alto.

Figure 3.11 Johnson and Sandage, 1955, The Astrophysical Journal, **126**, 326, The University of Chicago Press, Chicago.

Figure 3.12 Bower, Lucey and Ellis, 1992, Monthly Notices of the Royal Astronomical Society, **245**, 601, Blackwell Publishing, Oxford.

Figure 3.13 Kuntschner *et al.*, 2002, Monthly Notices of the Royal Astronomical Society, **337**, 172, Blackwell Publishing, Oxford.

Figure 3.14 Bridges, Hanes and Harris, 1991, The Astronomical Journal, **101**, 469, The University of Chicago Press, Chicago.

Figure 3.15 Halliday *et al.*, 2001, Monthly Notices of the Royal Astronomical Society, **326**, 473, Blackwell Publishing, Oxford.

Figure 3.16 Graham and Colless, 1997, Monthly Notices of the Royal Astronomical Society, **287**, 221, Blackwell Publishing, Oxford.

Figure 3.17 Bender, Burstein and Faber, 1992, The Astrophysical Journal, **399**, 462, The University of Chicago Press, Chicago.

Figure 3.21 Hibbard and van Gorkom, 1996, The Astronomical Journal, **111**, 655, The University of Chicago Press, Chicago.

Figure 3.22 Barnes and Hernquist, 1996, The Astrophysical Journal, **471**, 115, The University of Chicago Press, Chicago.

Figure 3.23 McClure and Dunlop, 2002, Monthly Notices of the Royal Astronomical Society, **331**, 795, Blackwell Publishing, Oxford.

Figure 4.1 Graham, 2001, The Astronomical Journal, **121**, 820, The University of Chicago Press, Chicago.

Figure 4.2 Valleé, 2002, The Astrophysical Journal, **566**, 261, The University of Chicago Press, Chicago.

Figure 4.3 Kennicutt and Edgar, 1986, The Astrophysical Journal, **300**, 132, The University of Chicago Press, Chicago.

Figure 4.4 Disney and Phillipps, 1983, Monthly Notices of the Royal Astronomical Society, **205**, 1253, Blackwell Publishing, Oxford.

Figure 4.5 O'Neil and Bothun, 2000, The Astrophysical Journal, **529**, 811, The University of Chicago Press, Chicago.

Figure 4.6 Giovanelli *et al.*, 1995, The Astronomical Journal, **110**, 1059, The University of Chicago Press, Chicago.

Figure 4.7 Sandage, Binggeli and Tammann, 1985, The Astronomical Journal, **90**, 1759, The University of Chicago Press, Chicago.

Figure 4.9 Gilmore and Reid, 1983, Monthly Notices of the Royal Astronomical Society, **202**, 1025, Blackwell Publishing, Oxford.

Figure 4.10 Malhotra, 1995, The Astrophysical Journal, **448**, 138, The University of Chicago Press, Chicago.

Figure 4.11 Worthey, 1994, The Astrophysical Journal Supplement Series, **95**, 107, The American Astronomical Society, Washington DC.

Figure 4.12 Zinn, 1985, The Astrophysical Journal, **293**, 424, The University of Chicago Press, Chicago.

Figure 4.13 Kroupa, Tout and Gilmore, 1993, Monthly Notices of the Royal Astronomical Society, **262**, 545, Blackwell Publishing, Oxford.

Figure 4.14 Kroupa, Tout and Gilmore, 1993, Monthly Notices of the Royal Astronomical Society, **262**, 545, Blackwell Publishing, Oxford.

Figure 4.15 Zwaan *et al.*, 2003, The Astronomical Journal, **125**, 2842, The University of Chicago Press, Chicago.

Figure 4.16 Diaz *et al.*, 1987, Monthly Notices of the Royal Astronomical Society, **226**, 19, Blackwell Publishing, Oxford.

Figure 4.17 Swaters, Sancisi and van der Hulst, 1997, The Astrophysical Journal, **491**, 140, The University of Chicago Press, Chicago.

Figure 4.18 Broeils and Rhee, 1997, A&A, **324**, 877.

Figure 4.19 Ferriere, 1998, The Astrophysical Journal, **497**, 759, The University of Chicago Press, Chicago.

Figure 4.20 Roy and Walsh, 1988, Monthly Notices of the Royal Astronomical Society, **288**, 715, Blackwell Publishing, Oxford.

Figure 4.21 Welsh, Sallmen and lallement, 2004, A&A, **414**, 261.

Figure 4.22 Haslam *et al.*, 1982, A&AS, **47**, 1.

Figure 4.23 Dame *et al.*, 1987, The Astrophysical Journal, **322**, 706.

Figure 4.24 Calzetti, Kinney and Storchi-Bergmann, 1994, The Astrophysical Journal, **429**, 582, The University of Chicago Press, Chicago.

Figure 4.25 Georgelin and Georgelin, 1976, A&A, **49**, 57.

Figure 4.26 Bally, O'Dell and McCaughrean, 2000, The Astronomical Journal, **119**, 2919, The University of Chicago Press, Chicago.

Figure 4.27 Kewley, Geller, Jansen and Dopita, 2002, The Astronomical Journal, **124**, 3135, The University of Chicago Press, Chicago.

Figure 4.28 Condon, Anderson and Helou, 1991, The Astrophysical Journal, **376**, 95, The University of Chicago Press, Chicago.

Figure 4.29 Kennicutt, 1998, The Astrophysical Journal, **498**, 541, The University of Chicago Press, Chicago.

Figure 4.30 Kennicutt, 1989, The Astrophysical Journal, **344**, 685, The University of Chicago Press, Chicago.

Figure 4.31 Matteucci, Romano and Molaro, 1999, A&A, **341**, 458.

Figure 4.32 Edvardsson *et al.*, 1993, A&A, **275**, 101.

Figure 4.33 Vila-Costas and Edmunds, 1992, Monthly Notices of the Royal Astronomical Society, **259**, 125, Blackwell Publishing, Oxford.

Figure 4.34 Begeman, 1989, A&A, **223**, 47.

Figure 4.35 Bosma, 1981, The Astronomical Journal, **86**, 1825, The University of Chicago Press, Chicago.

Figure 4.36 Begeman, Broeils and Sanders, 1991, Monthly Notices of the Royal Astronomical Society, **249**, 523, Blackwell Publishing, Oxford.

Figure 4.37 Stark *et al.*, 1992, The Astrophysical Journal Supplement Series, **79**, 77, The American Astronomical Society, Washington DC.

Figure 4.38 Dame *et al*, 1987, The Astrophysical Journal, **322**, 706, The University of Chicago Press, Chicago.

Figure 4.39 Begeman, 1989, A&A, **223**, 47.

Figure 4.40 Malhotra, Spergel, Rhoades and Li, 1996, The Astrophysical Journal, **473**, 687, The University of Chicago Press, Chicago.

Figure 4.41 Alcock *et al.*, 1997, The Astrophysical Journal, **479**, 119, The University of Chicago Press, Chicago.

Figure 4.42 Wang, 2002, Astronomische Nachrichten Supplementary Issue, **1**, 26, Wiley-VCH Verlag, Weinheim.

Figure 4.43 Schödel *et al.*, 2003, The Astrophysical Journal, **596**, 1015, The University of Chicago Press, Chicago.

Figure 4.44 Eisenhauer *et al.*, 2003, The Astrophysical Journal, **597**, L121, The University of Chicago Press, Chicago.

Figure 5.3 de Blok and Walter, 2000, The Astrophysical Journal, **537**, 95, The University of Chicago Press, Chicago.

Figure 5.4 Sandage, Freeman and Stokes, 1970, The Astrophysical Journal, **160**, 831, The University of Chicago Press, Chicago.

Figure 5.5 Richer and McCall, 1995, The Astrophysical Journal, **445**, 642, The University of Chicago Press, Chicago.

Figure 5.6 Bothun, Impey and Malin, 1988, The Astrophysical Journal, **330**, 634, The University of Chicago Press, Chicago.

Figure 5.7 Smecker-Hane *et al.*, 1996, ASP Conference Series, **98**, 32, Astronomical Society of the Pacific, San Francisco.

Figure 5.8 Mateo, 1997, ASP Conference Series, **116**, 259, Astronomical Society of the Pacific, San Francisco.

Figure 5.9 Aaronson and Mould, 1985, The Astrophysical Journal, **290**, 191, The University of Chicago Press, Chicago.

Figure 5.10 Skillman *et al.*, 2003, The Astrophysical Journal, **596**, 253, The University of Chicago Press, Chicago.

Figure 5.11 Putman *et al.*, 2003, The Astrophysical Journal,**586**, 170, The University of Chicago Press, Chicago.

Figure 5.12 Majewski *et al.*, 2003, The Astrophysical Journal, **599**, 1082, The University of Chicago Press, Chicago.

Figure 5.14 Bothun, Impey, Malin and Mould, 1987, The Astronomical Journal, **94**, 23, The University of Chicago Press, Chicago.

Figure 5.15 Sandage, Binggeli and Tammann, 1985, The Astronomical Journal, **90**, 1759, The University of Chicago Press, Chicago.

Figure 5.16 Smith, Driver and Phillipps, 1997, Monthly Notices of the Royal Astronomical Society, **287**, 415, Blackwell Publishing, Oxford.

Figure 5.17 Phillipps *et al.*, 1998, The Astrophysical Journal, **498**, L119, The University of Chicago Press, Chicago.

Figure 6.1 Capriotti, Foltz and Byard, 1980, The Astrophysical Journal, **241**, 903, The University of Chicago Press, Chicago.

Figure 6.2 Boyce *et al.*, 1998, Monthly Notices of the Royal Astronomical Society, **298**, 121, Blackwell Publishing, Oxford.

Figure 6.3 Francis *et al.*, 1991, The Astrophysical Journal, **373**, 465, The University of Chicago Press, Chicago.

Figure 6.4 Meyer *et al.*, 2001, Monthly Notices of the Royal Astronomical Society, **324**, 343, Blackwell Publishing, Oxford.

Figure 6.5 Türler *et al.*, 1999, A&AS, **134**, 89.

Figure 6.6 Brinks and Mundell, 1996, ASP Conference Series, **106**, 268, Astronomical Society of the Pacific, San Francisco.

Figure 6.7 Brinchmann *et al.*, 2004, Monthly Notices of the Royal Astronomical Society, **351**, 1151, Blackwell Publishing, Oxford.

Figure 6.8 Burns, Feigelson and Schreier, 1983, The Astrophysical Journal, **273**, 128, The University of Chicago Press, Chicago.

Figure 6.9 Hargrave and Ryle, 1974, Monthly Notices of the Royal Astronomical Society, **166**, 305, Blackwell Publishing, Oxford.

Figure 6.11 Biretta, Sparks and Macchetto, 1999, The Astrophysical Journal, **520**, 621, The University of Chicago Press, Chicago.

Figure 6.12 Urry and Padovani, 1995, Publications of the Astronomical Society of the Pacific, **107**, 803, University of Chicago Press, Chicago.

Figure 7.1 de Vaucouleurs, 1976, The Astrophysical Journal, **203**, 33, The University of Chicago Press, Chicago.

Figure 7.2 Hudson, 1993, Monthly Notices of the Royal Astronomical Society, **265**, 43, Blackwell Publishing, Oxford.

Figure 7.3 Kneib *et al.*, 1996, The Astrophysical Journal, **471**, 643, The University of Chicago Press, Chicago.

Figure 7.5 Miller and Owen, 2003, The Astronomical Journal, **125**, 2427, The University of Chicago Press, Chicago.

Figure 7.6 Rose *et al.*, 2002, The Astronomical Journal, **123**, 1216, The University of Chicago Press, Chicago.

Figure 7.7 Jenkins *et al.*, 2001, Monthly Notices of the Royal Astronomical Society, **321**, 372, Blackwell Publishing, Oxford.

Figure 7.9 Seldner, Seibers, Groth and Peebles, 1977, The Astronomical Journal, **82**, 249, The University of Chicago Press, Chicago.

Figure 7.10 Drinkwater *et al.*, 2000, A&A, **355**, 900.

Figure 7.11 Thorstensen *et al.*, 1989, The Astronomical Journal, **98**, 1143, The University of Chicago Press, Chicago.

Figure 7.12 Cruddace *et al.*, 2002, The Astrophysical Journal Supplement Series, **140**, 239, The American Astronomical Society, Washington DC.

Figure 7.13 Couch, Jurcevic and Boyle, 1993, Monthly Notices of the Royal Astronomical Society, **260**, 241, Blackwell Publishing, Oxford.

Figure 7.14 Phillipps *et al.*, 1978, Monthly Notices of the Royal Astronomical Society, **182**, 673, Blackwell Publishing, Oxford.

Figure 7.15 Bahcall and West, 1992, The Astrophysical Journal, **392**, 419, The University of Chicago Press, Chicago.

Figure 7.16 Hawkins *et al.*, 2003, Monthly Notices of the Royal Astronomical Society, **346**, 78, Blackwell Publishing, Oxford.

Figure 8.1 Izotov and Thuan, 2004, The Astrophysical Journal, **602**, 200, The University of Chicago Press, Chicago.

Figure 8.2 Driver *et al*, 1996, The Astrophysical Journal, **461**, 525, The University of Chicago Press, Chicago.

Figure 8.3 Tonry *et al.*, 2003, The Astrophysical Journal, **594**, 1, The University of Chicago Press, Chicago.

Figure 8.4 van Breugel *et al.*, 1999, The Astrophysical Journal, **518**, 61, The University of Chicago Press, Chicago.

Figure 8.5 Coleman, Wu and Weedman, 1980, The Astrophysical Journal Supplement Series, **43**, 393, The American Astronomical Society, Washington DC.

Figure 8.6 Fernandez-Soto, Yahil and Lanzetta, 1999, The Astrophysical Journal, **513**, 34, The University of Chicago Press, Chicago.

Figure 8.7 Weymann *et al.*, 1998, The Astrophysical Journal, **505**, 95, The University of Chicago Press, Chicago.

Figure 8.8 Roche *et al.*, 1997, Monthly Notices of the Royal Astronomical Society, **288**, 200, Blackwell Publishing, Oxford.

Figure 8.9 Liske *et al.*, 2003, Monthly Notices of the Royal Astronomical Society, **344**, 307, Blackwell Publishing, Oxford.

Figure 8.10 Croom *et al.*, 2004, Monthly Notices of the Royal Astronomical Society, **349**, 1397, Blackwell Publishing, Oxford.

Figure 8.11 Wright, 2001, The Astrophysical Journal, **553**, 538, The University of Chicago Press, Chicago.

Figure 8.12 Pettini, 2001, European Space Agency Conferences SP-460, European Space Agency, Paris.

Figure 8.13 Couch and Sharples, 1987, Monthly Notices of the Royal Astronomical Society, **229**, 423, Blackwell Publishing, Oxford.

Figure 8.14 Mann *et al.*, 2002, Monthly Notices of the Royal Astronomical Society, **332**, 549, Blackwell Publishing, Oxford.

Figure 8.15 Pettini, 2001, European Space Agency Conferences SP-460, European Space Agency, Paris.

Figure 8.16 Becker *et al.*, 2001, The Astronomical Journal, **122**, 2850, The University of Chicago Press, Chicago.

Figure A1 Fukigita, Shimasaku and Ichikawa, 1995, Publications of the Astronomical Society of the Pacific, **107**, 945, University of Chicago Press, Chicago.

Bibliography

There is a huge literature on the subject of galaxies. I list here a mixture of interesting background reading, texts at an accessible level and some more advanced texts on various topics.

History

Lonely Hearts of the Cosmos, Dennis Overbye, McMillan, London, 1991.
The Expanding Universe, Robert Smith, CUP, Cambridge, 1982.
The Realm of the Nebulae, Edwin P. Hubble, Yale UP, New Haven, Connecticut, 1936.
Galaxies (3rd Edition), Harlow Shapley, Harvard UP, Cambridge, MA, 1972.

General

Galaxies in the Universe, Linda Sparke and Jay Gallagher, CUP, Cambridge, 2000.
Galaxies and Galactic Structure, Debra M. Elmegreen, Prentice-Hall, Upper Sadler River, New Jersey, 1998.
Galaxies: Structure and Evolution, Roger J. Tayler, CUP, Cambridge, 1993.
Galactic Astronomy, James Binney and Michael Merrifield, Princeton UP, Princeton, NJ, 1998.

Morphology

Galaxy Morphology and Classification, Sidney van den Bergh, CUP, Cambridge, 1998.

The distance scale

Measuring the Universe: The Cosmological Distance Ladder, Stephen Webb, Springer, (Praxis), Chichester, 1999.

The Structure and Evolution of Galaxies   Steven Phillipps
© 2005 John Wiley & Sons, Ltd

Dynamics

Galactic Dynamics, James Binney and Scott Tremaine, Princeton UP, Princeton, NJ, 1987
   (Advanced).
Dynamics of Galaxies, Guiseppe Bertin, CUP, Cambridge, 2000 (Advanced).

Spirals

The Milky Way as a Galaxy, Gerard Gilmore, Ivan King and Pieter C. van der Kruit, University
   Science Books, Mill Valley, California, 1990.
The Andromeda Galaxy, Paul W. Hodge, Kluwer, 1992.

Interstellar medium

Physical Processes in the Interstellar Medium, Lyman Spitzer, Wiley, New York, 1978 (Classic).
The Dusty Universe, Aneurin Evans, Wiley, Chichester, 2000.

Nucleosynthesis

Nucleosynthesis and the Chemical Evolution of Galaxies, Bernard E.J. Pagel, CUP, Cambridge,
   1997.

Dwarfs

Galaxies of the Local Group, Sidney van den Bergh, CUP, Cambridge, 2000.

Active galaxies

An Introduction to Active Galactic Nuclei, Bradley M. Peterson, CUP, Cambridge, 1997.
Active Galactic Nuclei, Ian Robson, Wiley, Chichester, 1999.
High Energy Astrophysics, Volume 2, Stars, the Galaxy and the Interstellar Medium, Malcolm S.
   Longair, CUP, Cambridge, 1994.

The distribution of galaxies

Large-Scale Structures in the Universe, Anthony P. Fairall, Wiley, (Praxis), Chichester, 1998.
Modern Cosmology, Observations and Problems, Greg D. Bothun, Taylor & Francis, London,
   1998.
The Large Scale Structure of the Universe, P.J.E. Peebles, Princeton UP, Princeton, NJ, 1980
   (Advanced).

Cosmology

An Introduction to Modern Cosmology, Andrew Liddle, Wiley, Chichester, 2003.
An Introduction to the Science of Cosmology, David J. Raine and E.G. Thomas, IoP Publishing, Bristol, 2002.
Cosmological Physics, John A. Peacock, CUP, Cambridge, 1999 (Advanced).

Galaxy formation

Galaxy Formation, Malcolm S. Longair, Springer, Berlin, 1998.
The Road to Galaxy Formation, William C. Keel, Springer (Praxis), Chichester, 2002.

Review papers

Many excellent reviews on particular topics are also to be found in the pages of *Annual Reviews of Astronomy and Astrophysics*.

Web sites

The key web site for galaxy data and images is the NASA Extragalactic Data base at http://nedwww.ipac.caltech.edu.

Index

Printed and bound by CPI Group (UK) Ltd, Croydon, CR0 4YY

27/10/2024

14580378-0004